Irrigation Systems Engineering

nipa

New India Publishing Agency

Pitam Pura, New Delhi-110 034

Irrigation Systems Engineering

by

Dr. Balram Panigrahi
Professor and Head
Department of Soil and Water Conservation Engineering
Orissa University of Agriculture and Technology
Bhubhaneswar, Orissa, India - 751 003

New India Publishing Agency
Pitam Pura, New Delhi-110 034

Published by
Sumit Pal Jain *for*
New India Publishing Agency
101, Vikas Surya Plaza, CU Block, L.S.C. Mkt.,
Pitam Pura, New Delhi- 110 088, (India)
Phone: 011-27341717, Fax: 011-27341616
Mobile : 09717133558
E-mail: newindiapublishingagency@gmail.com
Web: www.bookfactoryindia.com

Irrigation Systems Engineering

ISBN : 978-93-80235-38-7

Composed and Designed by NIPA

Prof. D.P. Ray
Vice Chancellor

visit us at : ouat.ac.in
Tel : (0674) 2397700 (O), 2561606 (R)
Fax : 91+674-2397780
E-mail : vcouat@indiatimes.com
ouat_dproy@yahoo.co.in
Orissa University of Agriculture & Technology
Bhubhaneswar-751003

Foreword

The per capita availability of water is declining day by day. There is a large mismatch between the irrigation potential created and that actually utilized due to faulty management of irrigation water. The large scale investment and the environment related problems associated with major and medium irrigation projects have now ushered to pursue alternate avenues to increase the irrigation potential in the country. Since the large chunk of rainwater in the country is not properly utilized, it is felt to conserve rainwater to augment the water wealth of the country.

In-situ moisture conservation is the cheapest method of all moisture conservation methods, yet it cannot fully support water requirement of the crops. Artificial ground water recharge is very costly and cannot be afforded by many farmers. In this context, conservation of excess rainwater in farm ponds is highly essential. Considerable amount of research information are now available on the benefits of farm ponds in the rainfed areas and the ways to increase the use efficiency of stored water.

The book on **"Irrigation Systems Engineering"** authored by Dr. Balaram Panigrahi, Professor & Head, Department of Soil and Water Conservation Engineering, OUAT contains useful information on selection of site for construction of farm pond, design and construction of the farm pond, methods to reduce the losses and increase the storage capacity, efficient use of the harvested water, economic analysis of farm pond irrigation system and maintenance of farm pond in 13 Chapters. The book will serve as an invaluable resource for the students of Agriculture in general and Soil & Water Conservation Engineering and Agricultural Engineering in particular. It will also be useful for practicing engineers in the field of Agriculture and Civil Engineering, field extension and research workers, policy makers, teachers and all those interested in the conservation of soil and water and development of command areas.

I wish the Book all success.

D. P. Ray

Preface

This is a text book for agriculture and agricultural engineers and will be very much helpful for the beginning students in irrigation. It is designed to guide students from a basic knowledge of soil, mathematics, hydrologic and hydraulics to the state-of-the-art irrigation system design and management. Since major and medium irrigation projects are too costly and at the same time are not eco-friendly, the major thrust of research is now being imparted on low cost and easy to construct farm irrigation structures. The primary aim of the book is to design an optimum size small scale water harvesting structure which is the farm pond mostly used by the farmers in the farms. My goal is to present the principles and concepts of farm irrigation in a simple manner to maximize the students learning, understanding and motivation. The method and order of presentation have been carefully developed and classroom tested to make this book a useful and effective teaching tool. The book will not only be a helping tool to the students and teachers in agriculture and agricultural engineering but also to all the practicing engineers, agriculturists, soil conservationists and agricultural extension workers who deal directly or indirectly with water management and other associated farm development works. However, the book cannot be used for design of complex hydraulic structures including dams and reservoirs.

The book contains 23 solved problems, 238 short and long type questions, 42 tables, 55 figures and more than 138 references which will be immensely helpful to the students and design engineers. Several field experimental results have also been incorporated in the book at appropriate sections to make the book interesting for the readers.

The book contains 12 chapters and each chapter contains several sections and sub-sections. The title of the various chapters and the contents of each chapter are as follows.

Chapter 1

Farm Pond Irrigation and Rainfed Agriculture. This chapter deals with the different advantages of irrigation including its effect on enhancing the agricultural production and productivity. It also gives an overview of different disadvantages of irrigation including large investments incurred on irrigation sectors (particularly on major and medium projects), and the mismatch between the irrigation potential created and actually used. The present scenario of crop production under the rainfed ecosystem with constraints is discussed in brief. The scope of augmenting the production through rainwater harvesting in farm ponds in rainfed ecosystems are also discussed in this chapter.

Chapter 2

Types of Farm Ponds. The different classifications of farm ponds based on shape and size, method of construction and suitability to topographic situations are explained in this chapter. Lined and unlined ponds as well as shallow and deep farm ponds are also constructed in the farms and are presented in this chapter. The chapter also discusses the characteristic features of ponds suiting to local landscapes like sunken, barrage and diversion ponds. Finally the mode of selection of a particular type of pond in a farm is presented.

Chapter 3

Location and Site Selection. It is very important to locate a suitable site for construction of farm pond. The criteria for selection of suitable site are discussed. The runoff generating capacity of the catchment in which the pond is located for construction and the factors affecting the runoff of the catchment are also discussed in this chapter.

Chapter 4

Water Supply to Farm Pond. The two main sources of water supply to farm ponds are *in-situ* rainwater falling over the pond and runoff of the catchment entering the pond. The different climatic and physiographic factors influencing runoff of the catchment are discussed in this chapter. Runoff estimation by runoff coefficient, runoff percentage, Barlow's percentage, Strange's tables and curves, synthetic unit hydrograph method as well as rainfall-runoff modeling techniques are discussed. Methods to compute the runoff rate including rational and S.C.S. Curve Number method are also presented.

Chapter 5

Water Demand from Farm Pond. Harvested water of the farm pond is mostly used to meet the irrigation demand of the crops grown in the command area of the pond. Irrigation demand of the crops is dependent on the water requirement or crop evapotranspiration. Estimation of crop evapotranspiration by lysimeter study, field experimental plots, soil moisture depletion studies and water balance method has been discussed in this chapter. Climatological approaches for the estimation of reference evapotranspiration including the practical based pan evaporation method are also presented. Methods to estimate the actual crop evapotranspiration by selection of appropriate crop coefficient is also included in this chapter.

Chapter 6

Dimensions of Farm Pond: In this chapter, computations of various dimensions of farm pond like depth and side slope of the pond as well as various dimensions at the bottom, ground and top level have been presented. Design capacity/storage capacity of dugout and embankment farm pond has been set forth. Methods to find out the top water surface/water spread area and wetted surface area of farm pond has also been included. The chapter also includes a nomograph relating the depth, storage volume, top water surface area and wetted surface area for various sizes of farm ponds which will be helpful in computation of evaporation and seepage losses in farm pond.

Chapter 7

Optimum Size of Farm Pond. Description of different models including the water balance model of rice and dry land crop, water balance model of farm pond and irrigation management practices to simulate the size of farm pond have been included in this chapter. User friendly software developed by using Visual Basic 6.0 program has been presented to simulate the size of the farm pond for various types of cropping systems, irrigation management practices, types of soil and climatic conditions. A flow chart for simulation of pond size for rice-mustard cropping system has also been presented.

Chapter 8

Construction of Farm Pond. Methodology for the construction of farm pond, design of spillway with inlet and outlet structures, design of grassed waterways for safe disposal of excess water from the pond, design of earthen embankment including different cross sections are the main parts of this chapter. The seepage flow analysis through the earthen embankment has been included and procedures for the stability of embankment have been discussed.

Chapter 9

Losses of Water in Farm Pond. The two major losses of water in farm pond *i.e.* evaporation and seepage are discussed in this chapter. Various methods to compute the evaporation of farm pond including water budget, energy budget, mass transfer, empirical and pan evaporation methods are included. Techniques to reduce evaporation and seepage losses in farm pond are presented. Discussions on various sealants and their cost effectiveness are set forth. The chapter also includes siltation of farm pond and methods to reduce it.

Chapter 10

Efficient Use of Harvested Water of Farm Pond. This is the main chapter of the book. Several methods for quantifying irrigation requirements and schedules are presented. Techniques to decide the timing of irrigation to crops are discussed. Methods of irrigation and irrigation water measurement also form parts of the chapter. Different water lifting devices including indigenous methods and pumps are presented. Methods to increase the water productivity including land levelling and land grading are put forth. Results of water productivity of some field investigations are presented. Study of performance of farm irrigation system has also been included in this chapter.

Chapter 11

Economics of Farm Pond Irrigation System. This chapter, contains methods to compute the initial investment and annual operating cost of farm pond irrigation system. Economic analysis by present worth technique is discussed and procedures to estimate the economic parameters are covered. Results of economic analysis of some field studies are also presented.

Chapter 12

Farm Pond Maintenance. Maintenance of the farm pond against silting, control of aquatic weeds, maintenance of spillways and embankment of pond, reduction of pond bank erosion and cares to be taken while lining the farm ponds are discussed in this chapter.

Chapter 13

Other Sources of Irrigation. Canal Irrigation, Well Irrigation, Lift Irrigation, Pitcher Irrigation are discussed in this chapter.

Bhubhaneswar (Orissa) **(Balram Panigrahi)**
14 July, 2010

Contents

Chapter 1

Farm Pond Irrigation and Rainfed Agriculture

1.1 Introduction

Water is the basis of all forms of life. It supports life system not only to the plants but also to all other animals. The embryo from the conception till birth floats in the liquid medium. Cell, the basic unit of life, contains as high as 95- 98% water. Cellular metabolism is controlled by content of water. Majority of this water requirement of the crops is met from different forms of precipitation. However, when the water requirement is not fully met from precipitation, then growth of the crops is affected. This results in reduction of yield of the crops. The reduction in yield depends on the amount of scarcity of water requirement of the crops. Water requirement of crops depend on many factors. It depends on the type of crops, irrigation schedules practiced, soil texture, position of groundwater table below the effective root zone of the crop and management practice followed in crop cultivation. The pulses, oilseeds, vegetables which are also otherwise termed as dry land crops need less water compared to rice which is a water loving crops and perennial crops like sugarcane and banana.

Water requirement of crops is partly met by the upward flux of groundwater. The capillary rise of shallow water table in the form of upward flux helps in meeting the water requirement of crops to some extent. The upward flux depends on the type of the soil, effective root zone depth of the crop, depth of water table below effective root zone depth of the crop and potential gradient of soil moisture suction. The residual soil moisture helps in supporting water requirement of crops also. The heavy soil like clay has more moisture retention capacity than the light soils like sandy soil. The residual soil moisture held in clay soil is released slowly and supports soil moisture to the crops for a longer period than the sandy soil which has less soil moisture retention capacity. After meeting the water requirement of crops from all sources including different forms of precipitation, ground water contribution by capillary rise and residual soil moisture, if there is any shortage of water, then this shortage should be met from other sources which is called as irrigation.

1.2 Advantages of Irrigation

Introduction of irrigation has acted as a prime mover to the agricultural development programme by way of increasing and stabilizing the crop production. It has increased the cropping intensity and hence the total crop production and productivity of agricultural lands. It is needless to mention that the productivity of the agricultural lands have been enhanced by two to three times more than that of the rainfed lands. In future, the success of the second green revolution would become more feasible by increasing the irrigation potential and decreasing the gap between the irrigation potential created and that actually used. There is hardly any doubt in emphasizing the importance and advantages of irrigation during the periods of acute food shortages to feed the millions of starving population of the country. Even then, it is worthwhile to mention some of the tangent advantages of irrigation that are summarized as below (Sharma, 1987; Sharma and Sharma, 2002):

(i) *Agricultural Production*: Irrigation helps to add water to soil to supply the moisture essential for plant growth. It cools soil and atmosphere, thereby making more favourable environment for plant growth. It thus helps in increasing the total crop production and productivity, and hence, to attain self-sufficiency in food grains production.

(ii) *Domestic Water Supply*: Irrigation helps in augmenting the water supply in towns where water is available with great difficulty. It also provides water for other uses like swimming, recreational uses, bathing, cattle drinking etc.

(iii) *Communication*: Embankments of irrigation channels are used as pathways for facilitating communication to the villages. Many of the ancient civilizations have come up on the banks of the rivers and on the widen embankments of big canals.

(iv) *Inland Navigation*: Sometimes larger irrigation canals are used for navigation purposes. People move from place to place by boats and steamers in the larger irrigation canals. Some larger irrigation canals in Thailand are still now being used as inland navigation.

(v) *Industrial Uses*: Irrigation water is used to meet the water demand of industries, thus facilitating the industrial growth.

(vi) *Generation of Hydro-electric Power*: Cheaper power generation can be obtained on projects, primarily designed for irrigation alone. Canal falls can be used for power generation.

(vii) *Crop Insurance*: Irrigation helps to provide crop insurance against short duration droughts.

(viii) *Dilution of Salts*: Irrigation helps in washing out or diluting salts in the soil which is called as leaching. Thus, it helps in reclamation of saline soils.

1.3 Effect of Irrigation on Crop Yield

The expansion of irrigation has not only directly enabled yield increases of all the crops including rice, it has also enabled the farmers to use more advanced technology, field inputs including high yielding and hybrid seeds, more use of chemical fertilizers and pesticides and adoption of farm mechanization. About 60% of total agricultural production comes from the irrigated agriculture. Furthermore, the spread of irrigation has also enabled crop diversification (Hansen *et al.*, 1979). Crop yields have increased substantially and irrigated crops produce much more than the rainfed crops. Consequently, the food production has quadrupled during the last 50 years resulting in achieving the annual per capita availability of food grain of 200 kg in the country. Fig. 1.1 (Food and Agriculture Organisation, 1999) represents the influence of irrigation on yield of various crops. Data of Fig. 1.1 reveals that on an average, the yield of different crops under irrigated condition is 42% more than the yield of the rainfed crops. The principal crop rice has 70% enhancement of yield under the irrigated condition than the corresponding yield under the rainfed condition. It is worthwhile to mention here that the success of green revolution was mainly attributed to the expansion of irrigation and it was mainly the state Punjab which contributed significantly to the green revolution. Even today the states like Punjab, Haryana, U.P. etc. where the irrigation facility is more, the total production and productivity of different crops are also more.

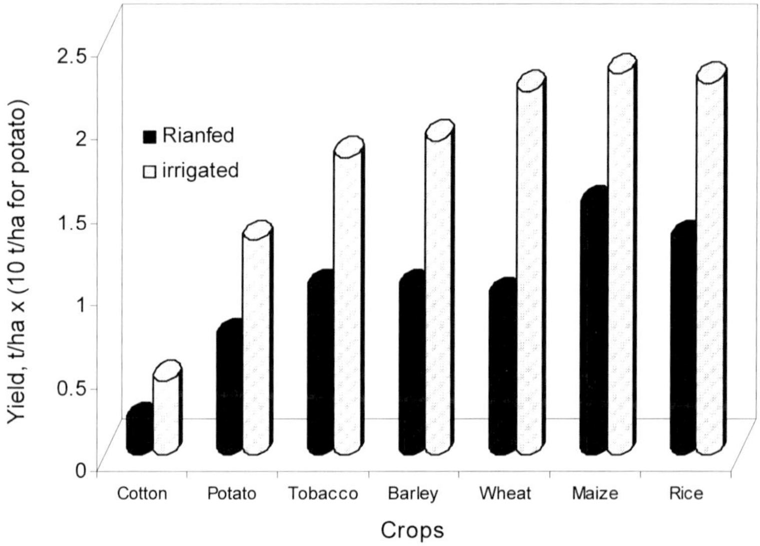

Fig. 1.1. Influence of irrigatoin on crop yield

1.4 Disadvantages of Irrigation

Besides many advantages, irrigation has also caused curse to many of the commands. It is obvious that major irrigation projects have less control. Canal irrigation has less efficiency compared to the well and other sources of irrigation. Irrigation projects, especially major irrigation projects cause many disadvantages as outlined below:

(i) Outbreak of Diseases

Canal irrigation through the major irrigation projects has less control and in general because of improper operation and maintenance causes swampy and marshy climatic conditions which in turns acts as breeding places of mosquitoes and flies. This causes outbreak of diseases like malaria.

(ii) Pollution Problems

Sometimes, farmers, especially in the head end of the canal system maintain deep ponding in paddy fields, which not only causes more losses of precious irrigation water but also decreases the yield of crop. It decreases the soil health, increases the iron toxicity in paddy field and thus decreases the yield. Besides, deep ponding in paddy field decreases the nitrogen and other fertilizer use efficiency in the crop fields, and causes groundwater pollution/contamination by increasing the nitrogen content in groundwater. Contaminated groundwater causes many diseases like anemia, cancer etc. Moreover, deep ponding in paddy field increases methane emission and thus creates environmental pollution. In irrigated areas, farmers use more agro chemicals, pesticides etc. which get washed from the fields, and enhances groundwater and soil pollution, Washing away of these chemicals from the fields ultimately reaches the rivers and seas and affects marine lives and affects biosphere.

(iii) Waterlogging and Salinity

Massive irrigation development in post plan irrigation periods in India has caused twin problems of waterlogging and salinity in almost all canal commands. This has not only threatened the capital investment but also the sustainability of irrigated agriculture has become an environmental concern too. In India, waterlogging in canal irrigated areas is increasing at an alarming rate. For example, in the Tungabhadra Project in Karnataka, more than 33000 ha has been reported to suffer from waterlogging and salinity and the areas are increasing at a rate of 600 ha/year (Lenka, 2001). Large areas in Haryana have been rendered waterlogged after the introduction of Bhakra canal irrigation. Rising water table in Gujarat has been observed due to flow of water in Mahi command. In eastern India, waterlogging is a serious problem. It is more serious in head reaches of canal irrigation system where over irrigation and faulty water management practices exist. The waterlogged saline soils are found to occur all over the country; about 8.5 M ha area has been salt-affected, of this 5.6 M ha is waterlogged saline area in the commands of irrigation projects and is commonly referred to as man-made or wet deserts (Balakrishnan *et al.,* 2001).

(iv) Increasing Flooding in High Water Table Areas

The continuous rise in water table in canal commands reduces the capacity of underground storage of water in the aquifers. There is reduction of ground water recharge in waterlogged tracts due to the non-availability of storage space in the aquifers beneath the tract. The reduced amount of natural recharge adds to rainfall excess, thereby enhancing the occurrence of floods.

(v) Sea Water Intrusion in Coastal Areas

Over exploitation of ground water near the sea coast creates imbalance between the sea water and contiguous ground water level: consequently the sea water intrudes the coastal aquifers and deteriorates the ground water quality. It also increases the salinity level of soil and water in the filed and thereby affects the crop yields.

(vi) Declining Ground Water Table

There is massive groundwater exploitation in many areas like Punjab, Haryana, Gujrat and some parts of Tamil Nadu in India. The heavy pumpage of groundwater in these areas has caused ground water mining which has caused decline of water level and thereby causing many wells to go dry in summer. Sometimes unplanned ground water pumpage causes landslide also.

1.5 Investment on Irrigation

Construction of major, medium and minor irrigation projects has enabled to create large irrigation potential in the country. The annual gross area under irrigation upto1951 (pre-plan period) from different sources was about 22.6 M ha. It rose to 54 M ha by 1983-84 which means that about 30% of the cropped area came under irrigation in the year 1983-84. It further rose to 81 M ha in the year 2002. Expansion of irrigated areas in the country has been possible with large scale investments on irrigation sector during successive plan periods. Beginning with an expenditure of Rs. 366 crores during the first plan, it rose to Rs. 10,786 crores during the sixth plan. During the seventh plan, an outlay of Rs. 17861 crores has been made on irrigation projects. Out of Rs. 17861 crores, major and medium projects shared Rs. 11556 crores and minor projects shared Rs. 6305 crores. The tenth plan had an outlay of Rs. 95743.42 crores for the irrigation sector. Expenditures incurred by the Government of India on irrigation sector during the successive years are presented in Fig. 1.2. The cost of creating per hectare of major and medium irrigation potential has increased from Rs. 1200 during the first plan to Rs.15018 during the seventh plan.

In minor irrigation projects, per hectare cost averaged Rs. 2452 till the fourth plan. It increased to Rs. 4574 and Rs. 7331 during sixth and seventh plans, respectively. Thus, the cost per hectare under major and medium irrigation projects during seventh plan has been 4.8 times to that of minor irrigation projects (Singh, 2001). Investment of such a large share of economy on major and medium irrigation sectors and comparing this huge investment with the returns obtained from irrigation is now questionable.

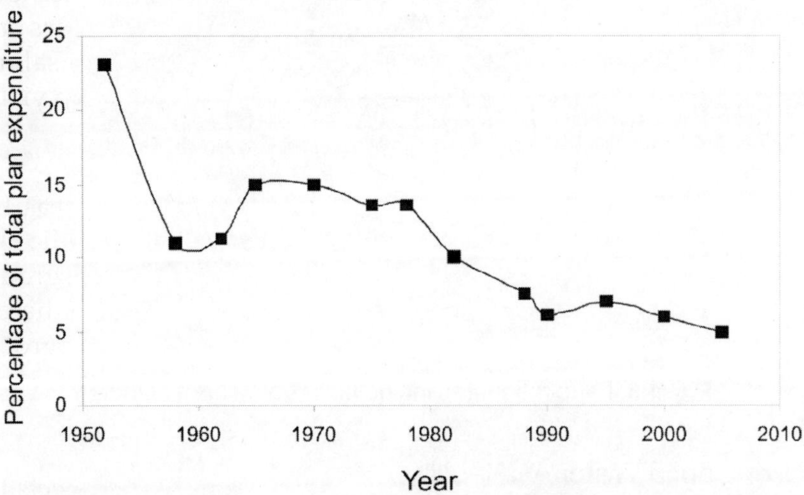

Fig. 1.2. Expenditure incurred on Irrigation projects during different years

1.6 Irrigation Potential Created and Actually Used

The ultimate irrigation potential created in India from different major, medium and minor schemes is 140 M ha. It is observed that there is a large mismatch between the irrigation potential created and that utilized in all the states of India. Up to the end of the sixth plan, the gap between the potential created and that utilized from major, medium and minor irrigation projects was about 7 M ha which rose to 19.70 M ha by the end of the tenth plan. The physical achievement on potential created and potential actually utilized till the end of the tenth plan was 109.10 and 90.42 M ha, respectively With an average irrigation intensity of 140%, actual net irrigated area up to the end of the tenth plan is expected to be 64.8 M ha which is only 45% of net sown area. Considering the tenth plan period alone, it is revealed that the target of irrigation potential was 15.15 M ha out of which only 10.36 M ha was utilized (4.79 M ha short of the potential created). Up to the end of the tenth plan, the country has achieved the irrigation potential of 78% of the ultimate value (Anonymous, 2007).

Fig. 1.3 represents the distribution of land resources including the irrigated area that is used in India. From this figure it is observed that about 27.4 % of the total geographical area of India are presently under irrigation.

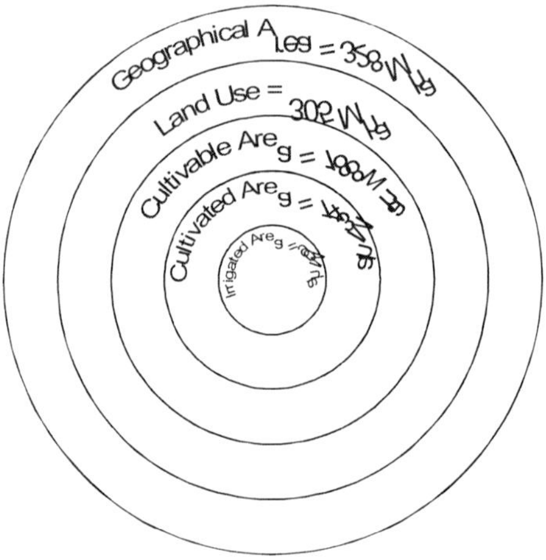

Fig. 1.3. Land resources including irrigated area of India

1.7 Per Capita Water Availability

The per capita water availability in the country is reducing progressively due to increasing population. The par capita annual water availability in 1955 was 5300 m^3 which has come down to 2100 m^3 as against the world average and Asian average of of 7420 and 3240 m^3, respectively. With the projected population growth, urbanization and industrialization, the annual average would come down to about 1400 m^3 by 2050 AD (Fig. 1.4, Tripathy, 2008).

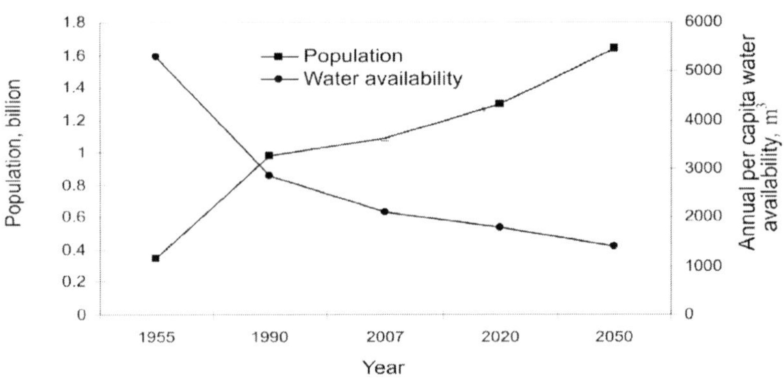

Fig. 1.4. Population and water avialability trends

1.8 Need and Scope of Water Resources Development

As the population of India continues to increase, the demand for food and fiber for the people will also increase. The country will have to feed about 1.3 billion people by 2020 AD and 1.64 billion people by 2050 A.D. which will require about 350 and 450 M tons of food grains by 2020 and 2050 AD, respectively. Considering the present level of annual food grains production, an additional food grain of 5 M tons/ year has to be produced in the country (Sahu, 2008). The average size of farm holdings is gradually decreasing, and there is no scope for adding any more area to crop cultivation since demand for other sectors especially for dwelling purposes are increasing rapidly. On the other hand, the water resources of the country is finite and the share of water for irrigation in agriculture is depleting at faster rate because of demand in other sectors like industrialization, community development, recreational uses etc. Considering an average annual rainfall of 1250 mm over 328 M ha area, about 400 M ha m of water is received by the country. About 200 M ha m of water enters the soil which constitutes soil moisture and ground water recharge. A considerable part of remaining 200 M ha m of water is flowing in the rivers inundating adjoining areas before entering seas and oceans. Annual water resources in the various river basins in the country is estimated to be about 187 M ha m. In addition, there is a dynamic rechargeable ground water resources whose potential has been estimated as 43.2 M ha m. This means that the total water available in the country is about 230 M ha m and per capita availability is about 2100 m^3/year. When the population will reach to 1.64 billion in 2050, according to U.N. agencies, per capita availability will decrease to 1400 m^3/year. The country will be water stressed by 2050 according to the World Bank and other agencies (Sivanappan, 2002).

In order to increase the irrigation potential a comprehensive control and management of rainwater is highly essential. Since the country has a vast rainwater resources and rainwater being the cheapest and easily available form of natural resource, it would be necessary to give importance of this part for effective harvest and utilization. It is to be noted that until unless effective planning on boosting agricultural production in the rainfed areas are not given due importance the country's economy would not be sound.

1.9 Rainfed Agriculture Ecosystem

Rainfed agriculture in India constitutes about 70% of the net sown area and contributes about 40% of the total food grains and supports 40% of the population in the country. Rainfed areas spread over almost all soil types which vary considerably in physical and chemical properties resulting in varied moisture situations. About 95% of the area under coarse cereals, 91% under pulses, 80% under oilseeds, 65% under cotton, and 53% under rice are rainfed (Bhatia and Singh, 2001). Rice is the major crop grown by most of the farmers in the country under rainfed situation. The area under rainfed rice in eastern India covers about 67% of the country's total rainfed rice area. But its contribution to the country's rice production is only about 50% (Ghildyal, 1989). Eastern regions of the country lag behind other regions in production and productivity of all crops including rice because most of the rainfed rice lands have inadequate water supply, uneven topography, small and fragmented land holdings along with serious soil and drainage problems.

Rainfed rice and drought usually go hand in hand for most of the areas in the eastern region. Although the region receives on an average annual rainfall of 1320 mm, yet there is considerable difference in spatial and temporal distribution (annual rainfall varying from 1090 to 2050 mm across the region). About 76% of the total annual rainfall occurs during the 4 months of the monsoon season (June - September). During this period, about 50% of the annual rainfall comes from a few intense storms which cause high runoff, soil and nutrient erosion (Pisharoty, 1990). The rainfall is highly erratic in distributions, frequently resulting in scarcity as well as excess moisture situation. Both these situations lead to unstable productivity and sustainability in crop production. A study by Panigrahi and Panda (2003) reveals that at 50% probability level the onset and withdrawal dates of the monsoon in the region would be 16th June and 3rd October with deviation of ± 5 days. Although the monsoon would be effective for 110 days, yet there would be two critical dry spells (CDS) of 12 days duration each occurring during this period out of which the second CDS would fall in the critical growth (reproductive) stage of a short duration rice of about 100-110 days which is grown to take maximum advantage of the monsoon rainfall in the region (Fig. 1.5). Rainfed rice will require supplemental irrigation during this period otherwise there will be severe yield reduction.

Assuming 40% of the rainfall contributing to runoff losses from a barren land, it is estimated that about 4.012 ton of runoff is produced from one hectare of land in monsoon season in the region and assuming every ton of runoff water carries a minimum of 2 kg fertile top soil from barren land, than 4.012 ton runoff will wash away about 8.024 tons of fertile top soil per hectare per year. This continuous process of soil erosion makes the soil impoverished resulting in barren lands. To make rainfed farming more sustainable, the pre-requisite is to conserve, develop and utiliise the two important natural resources i.e. soil and water more effectively and efficiently.

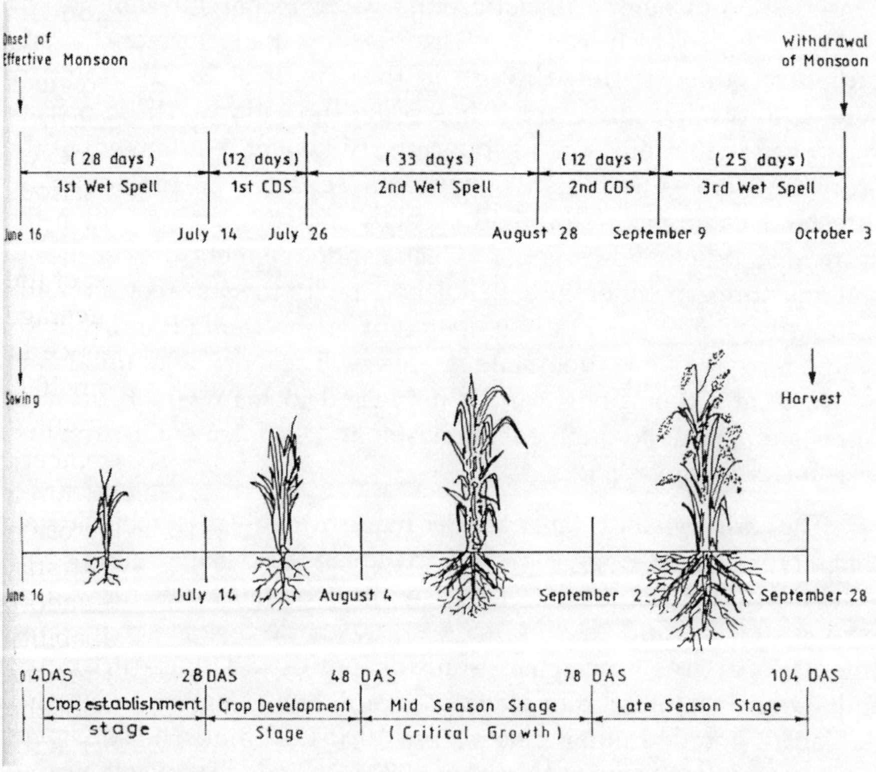

Fig. 1.5. Schematic presentation of growth stages of rice with occurrence of dry and wet spells

1.10 Scope of Rainwater Harvesting

It is reported that annually about 300-800 mm of seasonal rainfall is not used productively in the country. In the integrated watershed development program, there is ample scope for the surplus rainwater to be harvested in surface reservoirs or get recharged to the ground to augment the natural groundwater resource. There are good opportunities on in-situ conservation of rainwater at farm level with the excess water being taken out of the fields safely through community drainage channels and get it stored in suitable low cost structures including farm pond for direct use as supplemental irrigation to crops or for recharging the groundwater reservoir. The various methods of conserving soil and water include construction of mechanical soil conservation structures, *in situ* moisture conservation and artificial ground water recharge. However, it is to be marked that water conservation through artificial ground water recharge technique is a costly process which cannot be afforded by many farmers. In-situ moisture conservation though is the cheapest of all moisture conservation measures, yet it has some limitations. The moisture conserved in the soil cannot support the full water requirement of the crops and once the moisture contents deplete and reach certain threshold value below the field capacity moisture content, then the crops would demand irrigation which cannot be met from the stored soil moisture content of the soil. Further, in-situ moisture conservation technique cannot conserve 100 per cent of the rainfall and there is bound too have some runoff and soil losses. The inevitable runoff goes as waste and this runoff can be harvested in many small on-farm reservoirs/ponds to augment domestic and supplemental irrigation uses for growing crops.

The earlier water balance studies made by many researchers reveal that there is a great scope, especially in the semi-arid tropics and rainfed belts of the eastern region of the country for rainwater harvesting through farm pond reservoirs. A study made by the scientists at International Crop Research Institute for Semi-Arid Tropics (ICRISAT), India reveal that in high rainfall zones of Bhopal, Nagpur, Indore and Adilabad, annual rainfall varies from 1000-1200 mm. The soils have high water holding capacity of about 200 mm and mean water surplus for these regions ranged from 270-508 mm during the season. Water surplus ranged from 130 to 270 mm in 70% of the years and 230-475 mm in 50% of the years across the locations indicating vast scope to harvest the rainwater in farm ponds. Even, in the medium rainfall zones (annual rainfall around 700 mm) in Alfisols, the opportunities for water harvesting are > 70% of the years whereas for Vertisols it is less being about 50% (Wani *et al.,* 2003).

Panigrahi (2001) analysed the daily rainfall data for thirty years and forecasted that there would be three wet spells during the monsoon season in the eastern region varying from 25 to 33 days each and these three wet spells (total duration of three wet spells is 86 days) would contribute about 940 mm rainfall which is sufficient for cultivation of a suitable short duration rice. However, the distribution of rainfall is highly uneven and there are some periods when there is excess rainfall than the crop water requirement requiring drainage and some period when there is deficit especially during the critical growth stage (CGS) when the critical dry spell falls requiring supplemental irrigation. During the standard meteorological weeks from 35 to 37 which is the critical growth stage of rice, there is deficit supply of rainwater and for a getting good yield, irrigation requirement is sought (Fig. 1.6).

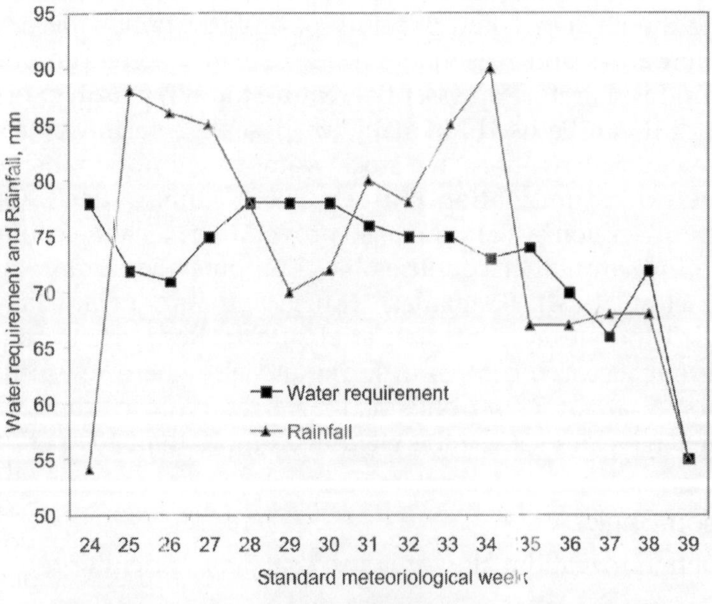

Fig. 1.6. Excess and dificit rainfall during different weeks

The technique of rainwater harvesting in small farm reservoirs/ ponds, depressions, valleys and low lying areas for meeting out the water demands of human beings, livestock and crops is not new and was practiced in India and many other countries since ancient days. The ponds prevailing in the southern parts of India give ample evidences of rainwater harvesting and utilizing the harvested water for multiple uses in low rainfall areas. Water stored in farm ponds can be used as handy sources for irrigation for a large number of crops. It can be used as multi source water for rearing fish in the reservoirs, growing crops in the command areas, meeting domestic uses including a viable source for drinking purposes. It also plays a vital role in integrated farming system thus enhancing the water-use efficiency and sustaining the livelihood security of the farmers.

Researches conducted in several areas of the rainfed farming ecosystems have revealed the feasibility of harvesting and storing excess rainwater in small farm reservoirs including dug out farm ponds and utiliising the same for providing supplemental irrigation to crops during critical dry spells with success and high pay off (Panigrahi, 2001; Panigrahi and Panda, 2003). An on-farm pond is one of the most appropriate technologies for water harvesting in northeast Thailand. The undulating landscapes, which dominate northeast Thailand, enhance the efficiency and usability of on-farm ponds, by providing catchment areas and reasonable head for water conveyance. An on-farm pond is therefore an essential feature for every farm in northeast Thailand. It can be used for many purposes e.g. farm water supply, supplementary irrigation, live stock water consumption, fish raising, and even flood mitigation (Suresh, 2002; Kumar, 1992). On farm reservoir irrigation systems are also proved to be techno-economically feasible in many other countries like Philipines (Guerra et al., 1990; Galang et al., 1994), Bangladesh (Musa and Kar, 1993; Islam et al., 1998) and Indonesia (Syamsiah et al., 1994). Researchers conducted at different research centres in India and elsewhere reveal that one irrigation of about 5 cm from small farm reservoir to crops at critical growth stage can increase the yield even more than 100 per cent.

Question Banks

Q.1. Define irrigation. How it is necessary for agriculture?

Q2. What are the advantages and disadvantages of irrigation?

Q3. Major and medium irrigation projects are both bone and curse to mankind. Justify.

Q4 How does irrigated agriculture differ from rainfed agriculture? How does irrigation affect production and productivity of crops?

Q5 With diagram, describe the land and water resources of the country.

Q6. How does the per capita water availability change with time?

Q7 There is enough scope for water resources development for irrigation in the country. How can these scopes be achieved?

Q8. What do you mean by rainfed agriculture? How it differs from irrigated agriculture?

Q9. How does the rainfall pattern affect the yield of crops in rainfed agriculture?

Q10. What is the importance of rainwater harvesting in rainfed agriculture and how this is achieved?

Q11. How is water harvesting through farm ponds helpful to enhance agricultural production?

Q12. Name some countries in the world where farm pond irrigation system is found to be viable for agriculture.

References

Anonymous, 2007. 11th Five Year Plan, Planning Commission Report, Government of India, New Delhi.

Balakrishnan, P; Lingappa, S; Shirahatti, M.S.; Kuligod, V.B, and Kulkarni, G.N. 2001. Reclamation of waterlogged saline areas in upper Krishna project command of Karnataka state, India. First Asian Regional Conference, Seoul, Korea (ICID): 55-61.

Bhatia, P.C. and Singh, H.P. 2001. Rainfed Agriculture: Research and Development Perspective. Yojana, January, 45 (1): 57-62.

Food and Agriculture Organisation. 1999. Irrigation in Asia in figures. Water Report No. 18, Rome, pp. 91-106

Galang, A.L., Bhuiyan, S.I. and Hunt, E.D. 1994. Identification of potential areas for use of the on-farm rreservoir system for drought alleviation. In: On-farm Reservoir Systems for Rainfed Ricelandss. Edited by S. I. Bhuiyan. International Rice Research Institute, Manila, Philippines, pp. 21-34

Ghildyal, B.P. 1989. Rice production in eastern India-issues in rice research. Journal of Agriculture Issues. Vol. 1: 79-91.

Guerra, L.C. Watson, P.G. and Bhuiyan, S.I. 1990. Hydrological analysis of farm reservoirs in rainfed areas. Agricultural Water Management, Vol. 17(4): 351-366.

Hansen, V.E., Israelsen, O.W. and Stringham, G.E. 1979. Irrigation Principles and Practices. John Wiley and Sons Inc., New York.

Islam, Md. T., Saleh, A.F.M. and Bhuiyan, S.I. 1998. Agro-hydrologic and economic analysis of on-farm reservoirs for drought alleviation in rained ricelands of northwest Bangladesh. Rural and Environmental Engineering, Japanese Society of Irrigation, Drainage and Reclamation Engineering, Vol. 35, 15-26.

Kumar, D. 1992. Fish Culture in Undrainable Ponds: A Manual for Extension. FAO Fisheries Technical Paper 325, FAO.

Lenka, D. 2001. Irrigation and Drainage. Kalyani Publishers, New Delhi.

Musa, A.M and Kar, N.K. 1993. Harvesting rainwater for crop production in Barind tract. On-Farm Research Division, Bangladesh Agricultural Research Institute, Gazipur, Bangladesh: 12-15.

Panigrahi, B. 2001. Water Balance Simulation for Optimum Design of On-Farm Reservoir in Rainfed Farming System. Ph.D. Thesis, Indian Institute of Technology, Kharagpur.

Panigrahi, B; and Panda, S. N. 2003. Optimal Sizing of On-Farm Reservoir for Supplemental Irrigation. Journal of Irrigation and Drainage Engineering, ASCE, Vol. 129 (2): 117-128

Pisharoty, P.R. 1990. Indian rainfall and water conservation. Proceedings of All India Seminar on Modern Techniques of Rainwater Harvesting, Water Conservation and Artificial Recharge for Drinking Water, Directorate of Groundwater Surveys and Development Agency, Government of Maharashtra, India, 50-53.

Sahu, G.C. 2008. Interlinking of Rivers for National Integration. Technical Annual of Institution of Enginers (India), Orissa State Centre: 40-49.

Sharma, R.K. and Sharma, T.K. 2002. Irrigation Engineering including Hydrology, S. Chand and Company Ltd. New Delhi.

Sharma, R.K., 1987. A Text book of Hydrology and Water Resources. Dhanpat Rai and Sons, Delhi.

Singh, S.R. 2001. Management of Land and Water Resources for Increasing Agricultural Productivity. Theme Paper Presented in the Annual Convention of Agricultural Engineers (Indian Society of Agricultural Engineers), Bhubaneswar.

Sivanappan, R.K. 2002. Expansion of Storage Capacity. Critical Survey of Indian Agriculture, The Hindu.

Suresh, R. 2002. Soil and Water Conservation Engineering. 2nd ed., Standard Publishers Distributors.

Syamsiah, I., Fagi, S. A. M. and Bhuiyan, S. I., 1994. Collecting and conserving rainwater to alleviate drought in rainfed ricelands of Indonesia. *In:* On-farm Reservoir Systems for Rainfed Ricelands. Edited by S.I. Bhuiyan. International Rice Research Institute, Manila, Philippines, pp. 142-152.

Tripathy, L.K. 2008. Sustainable Water Management. Technical Annual of Institution of Engineers (India), Orissa State Centre: 55-60.

Wani, S.P; Pathak, P; Sreedevi, T.K; Singh, H.P. and Singh, P., 2003. Efficient management of rainwater for increased crop productivity and ground water recharge in Asia. Comprehensive Assessment of Water Management in Agriculture Series, Water Productivity in Agriculture: Limits and Opportunities for Improvement. Book Edited by J.W.Kijne, R. Barker and D. Molden. CABI Publishing, pp. 352.

Chapter 2

Types of Farm Ponds

2.1 Introduction

Farm ponds are the most commonly used source of ex-situ water harvesting structures. It has wide adaptability in farmers' fields and used as handy source for water supply to crop fields. It has been used since time immemorial not only as an assured source of water supply to crops but also to meet demands in other sectors. However, these ponds have wide variations in shape and size as per the method of construction and their suitability for different topographic conditions. Most of the ponds, initially constructed in fields have regular shape. However, as time progresses and because of the poor or no repair and maintenance, these ponds loose their regular shape and become irregular.

2.2 General Classification of Pond

Followings are the general classification of farm ponds.

(i) Dugout pond

(ii) Surface pond

(iii) Spring or creek fed pond

(iv) Off -stream storage pond

(v) Seepage pond

(vi) Pump-fed pond and

(vii) Embankment pond

(i) Dugout Pond

Dugout ponds sometimes called as excavated ponds are so called because they are constructed in the ground by digging soil. They are generally constructed in flat topography. The dugout soil is used in construction of embankments of the ponds. These ponds collect both in-situ rainwater and surface runoff from the adjacent catchments. The harvested water is lifted for irrigation in command areas by means of pumps. Excavated ponds are designed with a regular shape either square or rectangular. There may be circular ponds. Out of the above three, the circular ponds have the highest storage capacity and least perimeter for a given surface area and side slopes (Table 2.1). Fig. 2.1 shows geometric features of farm ponds of different shapes with a given surface area of 900 m^2 for each case and having depth of 2 m in each case. However, it is difficult to construct such type of ponds. The square and rectangular types of ponds are therefore in common use. For the same pond size, the total dike length increases as the pond shape deviates from square to rectangular and as the shape is more elongated, the perimeter length increases more. At the same time, the construction cost also increases (Belgaumi et al., 1997). For an illustration, the dimensions and perimeter length (length of dikes) for different shapes of farm ponds having same size i.e. 400 m^2 are presented in Table 2.2 below.

Table 2.1. Perimeter and storage capacity of different shapes of farm ponds with given surface area (900 m^2) and depth (2 m).

Shape of pond	Side slope	Perimeter, m	Storage capacity, m^3
Circular	1.5:1	106	1499
Square	1.5:1	120	1464
Rectangular	1.5:1	122	1458

Table 2.2. Dimensions and perimeter of different shapes of 400 m^2 farm pond

Pond shape	Length: Width	Width, m	Length, m	Perimeter (length of dykes, m)
Square	1:1	20.00	20.00	80.00
Rectangle	2:1	14.14	28.28	84.85
	3:1	11.55	34.64	92.37
	4:1	10.00	40.00	100.00

Because of low construction cost, the square ponds are particularly useful as smaller upto 400 m^2. For ponds larger than 400 m^2 size, rectangular ponds can be constructed.

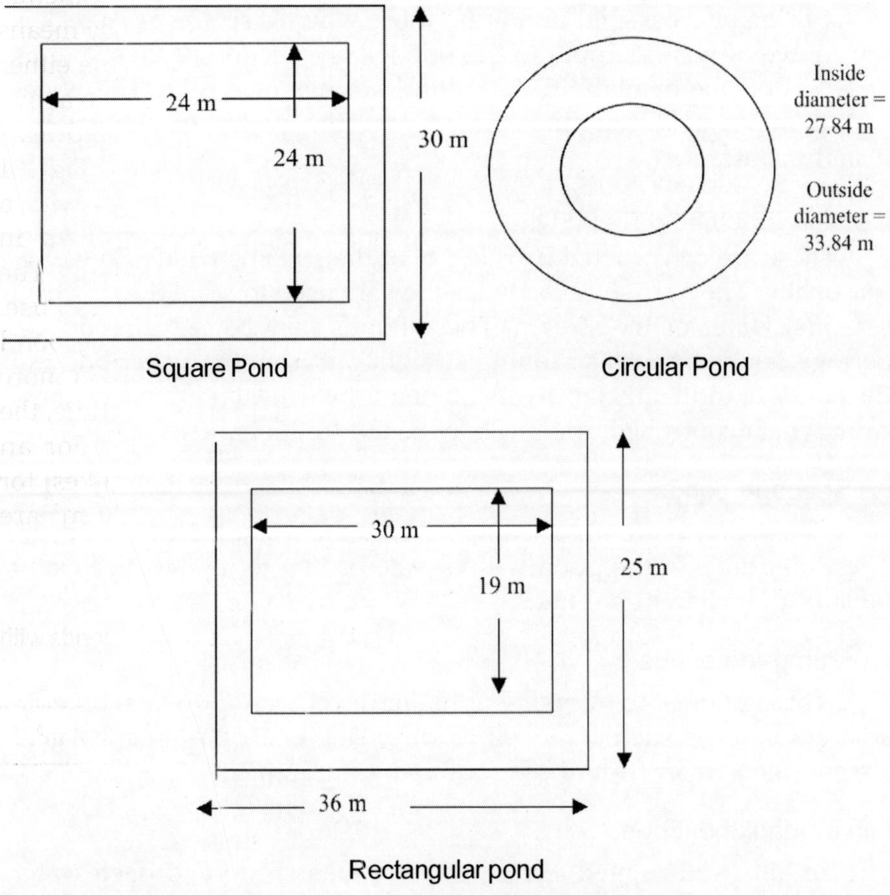

Fig. 2.1. Geometric features of different shape farm ponds

(ii) Surface Pond

Surface ponds also called as excavated cum embankment type pond are the most common types of farm ponds. They are partly constructed in ground where there is some depression. The dugout soil is utilized for the construction of the embankments of the ponds so that the pond can store maximum amount of water. These ponds are mostly fed by surface runoff although they catch in-situ rainfall and at times called as rainfed ponds. Normally they get no supply during the dry season. Water from these ponds flows out by gravity till the water level in the pond reaches the ground level and then the rest amount of stored water is pumped out. Selection of site of these ponds is therefore very important.

(iii) Spring or creek fed ponds

Spring or creeks fed ponds are those where a spring or a creek is the source of water supply to the pond. Construction of these ponds depends on the availability of natural springs or creeks. The water supply may vary throughout the year but the quality of the water is usually constant.

(iv) Off-stream storage ponds

These are constructed by the side of the streams which flow only seasonally. They are so called since they are used to store the diverted seasonal water of the stream. These ponds may be fed directly i.e. barrage ponds by water running straight out from the water body to the ponds or indirectly i.e. diversion ponds by water entering a channel from which controlled amount may be fed to the ponds.

(v) Seepage ponds

These ponds are supplied water from the water table by seepage. The water table in the pond will vary according to the level of water table in the adjacent aquifer.

(vi) Pump-fed ponds

These ponds are normally at higher level than the water level in adjacent sources such as a well, spring, lake or irrigation canal and are supplied water from these sources by pumping.

(vii) Embankment pond

In hilly and rugged terrain with frequent wide and deep water courses, embankment type ponds are also constructed.

2.3 Lined and Unlined Ponds

In addition to the above types, ponds may be lined or unlined types. Unlined ponds have no linings and are otherwise called as earthen ponds. They have high seepage losses and hence their storage capacity decreases with time after cessation of rainfall. The rate of seepage losses gradually decreases with deposition of silt in the beds and sides of the ponds. On the other hand, the lined ponds have low seepage rate and high storage capacity compared to the unlined ponds. The lined ponds are named depending on the types of lining. Some of the commonly used lined ponds are plastic lined pond, bitumen lined pond, clay lined pond, brick lined pond and cement-concrete lined pond (Panigrahi, 2001).

Some more types of ponds are walled ponds, drainable ponds, undrainable ponds, pump-drained ponds, cut and fill ponds etc.

Depending on the way the ponds fits in with the features of the local landscape, ponds can be conveniently grouped into three basic types i.e. sunken pond, barrage pond and diversion pond (Coche *et al.*, 1995).

2.4 Sunken Ponds

Following features are associated with sunken pond.

- The floor of the pond is generally below the level of the surrounding.
- The pond is directly fed by groundwater, rainfall and/or surface runoff. It can also be fed by pumping but is not normally supplemented.
- It is undrainable or partially drainable having been built either as dug-out pond or sometimes with additional embankments to increase the depth.
- They are created in the bottom of the valley by building a dam across the lower end of the valley. They may be built in a series down the valley as shown in Fig. 2.2.

2.5 Barrage Ponds

Barrage ponds have the following features:

- It is drainable through the old river bed.
- It may have diversion canal or without. If large floods are present, the excess water is normally diverted around one side of the pond

through a diversion canal. The pond water level then is controlled through water intake. Water enters directly to the pond from a spring, reservoir, lake etc. through a point called as inlet and flows out at a point called as outlet.

- To protect the pond dike from floods, a spillway is built (Fig.2.3).

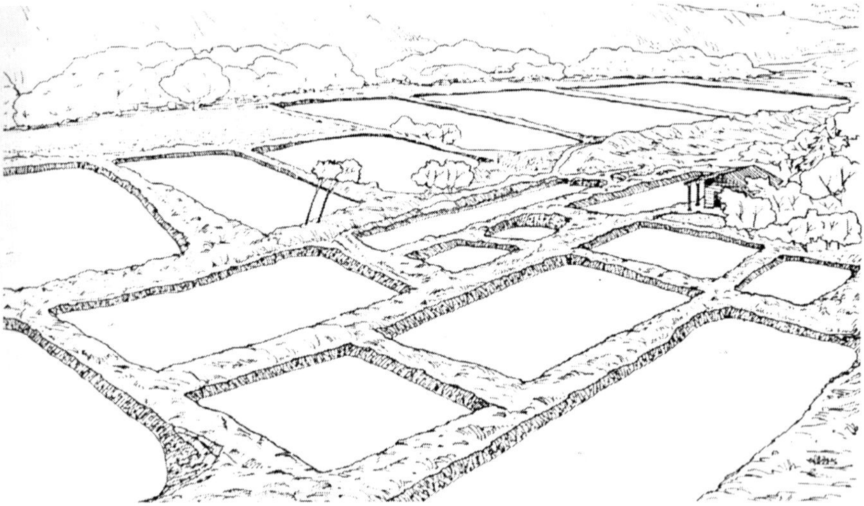

Fig. 2.2. Sunken ponds built on the bottom of a valley

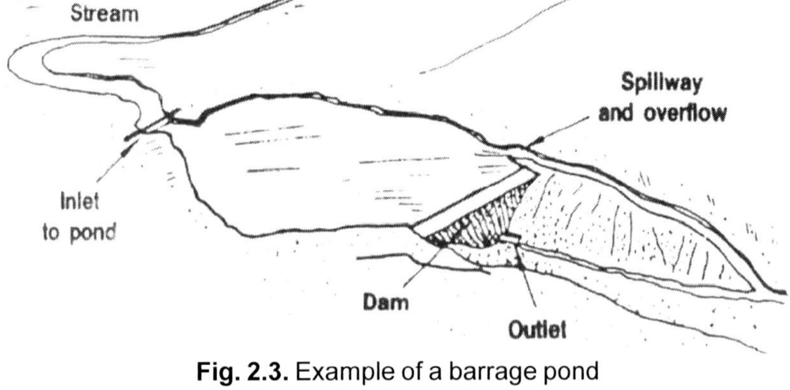

Fig. 2.3. Example of a barrage pond

2.6 Diversion Ponds

The features of the diversion ponds are:

- The diversion ponds are fed directly by gravity or by pumping through a diversion canal from a spring, lake or reservoir. The water flow is controlled through a water intake. There is an inlet and outlet for each pond.
- They can be constructed either in sloping ground as a cut and fill pond or in a flat ground as a four dike embankment pond sometimes called as paddy pond.
- It is usually drainable through a drainage canal.

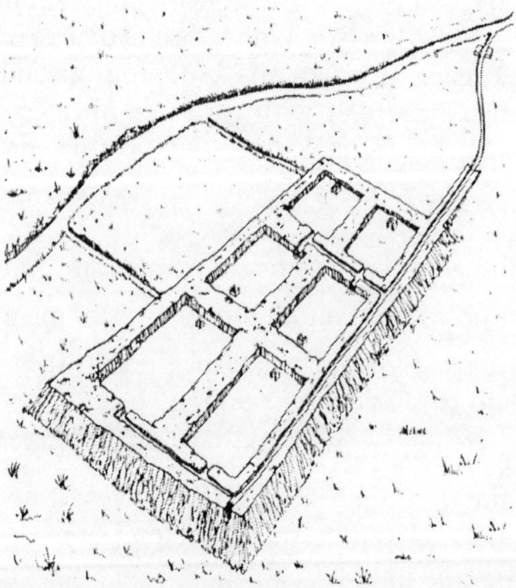

Fig. 2.4. Example of diversion ponds built in flat ground

It is important to note that the choice of a particular type of pond largely depends on the kind of water supply available and on the existing topography of the site selected. For an example if the valley bottom is 50 to 100 m wide, barrage ponds might be appropriate and when the valley bottom is more than 100 m wide, diversion ponds may be built. However, when there are choices of several types of ponds, one should give highest priority to diversion ponds fed by gravity and lowest priority to barrage ponds in flooding areas requiring large diversion canals (Coche *et al.*, 1995).

2.7 Shallow and Deep Ponds

Depending on the depth of water, the pond may be classified as shallow or deep pond. Except in some barrage ponds built on streams with steep longitudinal slope, fish ponds are generally shallow. Their maximum water depth is limited to 1.5 m. Their shallowest area should be at least 0.5 m deep to limit the growth of aquatic plants. The water depth in small ponds in many of the rural areas varies from 0.5 to 1 m. It is to be noted that the deeper ponds are much more expensive to build because of increasing exposed area and volume of earth works. Furthermore, the lifting of the excavated earth materials becomes difficult and requires more labour works and hence wages. However, it has some advantages too. In dry regions with more evaporation losses, deeper ponds are useful to reduce the evaporation loss and enable the pond users to have enough storage to meet the fish, domestic and agricultural uses. Even in the cold regions where it may be necessary to provide the fish with a refuge in deep ponds, warmer water are useful during the cold weather (Verma *et al.*, 1984).

Question Banks

Q1. What are the general classifications of farm ponds?

Q2. Write short notes on the following types of farm ponds.

(i) Dugout pond (ii) Embankment pond (iii) Sunken pond (iv) Barrage pond (v) Diversion pond.

Q3. Differentiate between the following types of farm ponds.

(i)Dugout pond and surface pond (ii) Lined pond and unlined pond (iii) Barrage pond and diversion pond

Q4. Which shape of dugout pond has the highest storage capacity and least perimeter?

Q5. Draw the geometric shape of a square size dugout pond having 1600 m^2 top surface area, 1:1 side slope and 2 m depth.

Q6. Draw the geometric shape of a rectangular dugout pond which has top surface area 1000 m^2, length:width is 2:1, side slope is 1:1 and depth is 2m.

Q7. What is the most common type of farm pond? How is it constructed?

Q8. How is the water lifted for irrigation from a farm pond?

Q9. Name any five types of lined farm ponds.

Q10. What are the differences between the lined and unlined pond?

Q11. Draw the diagram of a sunken pond built on the bottom of a valley.

Q12. Draw the diagram of a barrage pond and explain its characteristic features.

Q13. What are the factors on which the selection of a type of farm pond largely depends on? Enumerate with suitable example.

References

Belgaumi, M.I., Itnal, C.J. and Radder, C.D. 1997. Farm Ponds. Technical series 4, Publication Centrte, University of Agricultural Sciences, Dharwad, pp. 47.

Coche, A.G., Muir, J.F. and Laughlin, T. 1995. Simple Methods for Aquaculture Pond Construction for Freshwater Fish Culture Building Ponds, FAO of the United Nations, Rome, pp. 355.

Panigrahi, B. 2001. Water Balance Simulation for Optimum Design of On-Farm Reservoir in Rainfed Farming System. Ph.D. Thesis, Indian Institute of Technology, Kharagpur.

Verma , H.N., Prihar, S.S. and Singh, R. 1984. Feasibiliity of storage of runoff in dugout ponds and its use for supplemental irrigation. Indian Journal of Soil Conservation. Vol. 12(1): 31-36.

Chapter 3

Location and Site Selection

3.1 Introduction

Different types of farm ponds with their characteristics features are discussed in this Chapter. It is observed that farm ponds can be constructed at different sites starting from the plain areas with relatively flat slopes to valley terrain having steep slopes. The choice of a particular type of pond largely depends on the kind of water supply available and on the existing topography of the site selected. In general for valleys with bottom width less than 100 m, barrage ponds may work well whereas valleys with bottom width more than 100 m, diversion ponds find suitable.

3.2 Collection of Basic Information

Basic information on different data relating to the soil, climate, plant, topography, groundwater table, catchment characteristics, and livestock populations are to be collected for the area where pond has to be constructed. Estimation of water supply from different sources and water demand form the pond for various purposes are to be done and a pond water balance study needs to be done before designing a pond in a particular place.

3.3 Location of Farm Pond

Farm pond should be located in such a place in a farm that it can supply irrigation to all the crops in the farm. The pond should also get sufficient runoff from all the points in the farm. The most important criteria for selecting the location of the pond are the selection of catchment which conveys runoff water to the pond. Too big a catchment may bring enough runoff which may not be accommodated in a small size pond whereas too small catchment may not supply adequate amount of runoff to get stored in the pond. The runoff generating capacity of a catchment depends on a number of factors such as rainfall and physiographic factors (Sharma and Sharma, 2002). The rainfall factors include

(i) Amount of rainfall
(ii) Intensity, duration and frequency of rainfall
(iii) Distribution of rainfall and
(iv) Direction of storm movement.

The physiographic factors consist of both the watershed characteristics and channel characteristics. The watershed characteristics include:

(i) Size
(ii) Shape
(iii) Slope
(iv) Orientation
(v) Land use
(vi) Infiltration characteristics
(vii) Soil texture and
(viii) Topography.

Channel characteristics relate mostly to hydraulic properties of the channel that governs the water flow. These factors include:

(i) Size
(ii) Slope

(iii) Cross section and
(iv) Roughness coefficient of channel

The land use factors governing runoff include:

(i) Type of cultivation
(ii) Soil cover
(iii) Methods of soil and water conservation practices followed and
(iv) Crop management practices followed.

Based on the above mentioned several factors, the quantity of runoff varies considerably. Studies at different research centers reveal that the quantity of runoff varies from 10 to 55 % of the rainfall depending on the type of soil, slope and management practice. A study conducted by Sastry et al. (1985) reveals that sandy loam soil with 10-15% slope yields 55% rainfall as runoff whereas red loamy soil with slope 2-5% yields only 10% rainfall as runoff. Table 3.1 summarises the estimated water yield of various catchments for different soils and slopes at different locations in India (Sastry *et al.*, 1985).

Table 3.1. Estimated water yield into farm pond from different catchments

Region	Soil types	Annual rainfall, mm	Land slope, %	Estimated water yield, % of rainfall
Dehradun, Uttarakhand	Silty loam	1600	2-4	16.5
Bellary, Karnataka	Deep black	508	1_2	20.0
Shiwalik region, M.P.	Sandy loam	1150	10-15	55.0
Anantpur, A.P.	Shallow red	540	3-5	20.0
Hyderabad, A.P.	Red sandy loam	770	2_5	10,0
Bangalore, Karnataka	Red soil	830	2-3	20.0
Karnal, Haryana	Alkali	650	0-2	40.0
Chitradurga, Karnataka	Red loam	612	5-10	10.0
Bunga, Haryana	Silty clay loam to sandy loam	1116	10-12	50.0

Detail discussions on factors affecting the runoff of a catchment are mentioned in Chapter 4.

3.4 Selection of Site for Farm Pond

The selection of suitable site for locating a farm pond should begin with preliminary studies. A topographical map of the study area helps in identifying the suitable site for pond construction. In general there should have some catchment area above the pond and some command/service area below the pond. The runoff generated from the catchment should be harvested in the pond which can be reused as supplemental irrigation to crops grown in the same catchment area or in the command/service areas (Belgaumi *et al.*, 1997). Fig. 3.1 shows such a site where pond is located in a farm.

Fig. 3.1. An ideal site for construction of farm pond

Following points are to be kept in mind before selecting a site for pond construction.

(i) The site should be easily accessible.

(ii) The site should be as far as possible situated in the lowest depression which will require less earthwork for construction and collects more runoff.

(iii) As far as possible the site should not have loose, porous and coarse textured soil which is prone to more seepage losses causing reduced water storage of the pond.

(iv) The pond should have a natural water way or drainage channel to dispose off the excess storage of the pond quickly.

(v) The pond should be located in such a place that it is easily accessible for the live stocks to use the water of the pond.

(vi) It should be located in such a place that the water can be conveyed easily to all the parts of the farm area.

(vii) It should be away from the drainage lines of the sewerage washes, farm stead wastes, mine dumps and other similar things so that pond water is not contaminated.

(viii) Good command area with level land requiring little or no land shaping should be available near the pond.

(ix) The catchment of the pond should be so selected that the runoff from the catchment carries very less silt to the pond which will increase the storage capacity and life span of the pond.

(x) Finally, there should not have any power cable lines and other buried pipe lines below the pond site where it is constructed.

Question Banks

Q1. What is the basic information to be collected for site selection of a farm pond?

Q2. What are the factors on which the location of farm pond depends on?

Q3. Briefly describe the different factors which influence the runoff generating capacity of a catchment.

Q4. What are the different rainfall and physiographic factors that affect the runoff of a catchment?

Q5. How does a topographical map of an area help in selection of farm pond site?

Q6. What are the different points need to be considered for selecting site of a farm pond?

Q7. A site has loose and porous soil. Justify whether a farm pond should be constructed at this site.

References

Belgaumi, M.I., Itnal, C.J. and Radder, C.D. 1997. Farm Ponds, Technical series 4, Publication Centrte, University of Agricultural Sciences, Dharwad, pp. 47.

Sastry, G., Patnaik, U.S. and Dhruba Narayana, V.V. 1985. Proceedings of National on Soil Conservation and Watershed Management, New Delhi.

Sharma, R.K. and Sharma, T.K. 2002. Irrigation Engineering including Hydrology, S. Chand and Company Ltd. New Delhi.

Chapter 4

Water Supply to Farm Pond

4.1 Introduction

A farm pond is basically small scale water harvesting and water recycling structure. It harvests water mainly in the rainy season when there is abundant rainfall producing sufficient runoff and the harvested water may be used for single or multiple cropping. When there is deficient supply of water to the crops it may supplement as supplemental irrigation. The leftover water in the pond can be used for growing another crop in winter when there is no or less rainfall. Ij addition, the stored water in the pond can be used for domestic and other uses and rearing fish. Thus it serves as viable medium for integrated farming system in rainfed farming ecosystem. Before designing a pond in a given location, it is very important to estimate the total quantity of water supply to the pond. The two main sources of water supply to pond are (i) *in situ* rainfall and (ii) runoff (some times called as discharge) generated from the catchment and conveyed to the pond. Some times there is groundwater contribution to the pond when the pond is unlined and the underlying groundwater table with reference to the ground level is at lower elevation than the depth of water in the pond such that there is a hydraulic gradient in the direction of underlying groundwater level to the water level in the pond.

4.2 *In-Situ* Rainwater Estimation

The volume of rainwater conserved in a pond depends on the quantity of rainfall falling over the open surface area of the pond and the top or open surface area of the pond. If the quantum of rainfall falling over a given open surface area of the pond is more, then the pond will harvest more water and vice versa. Similarly, a given amount of rainfall falling over a pond will contribute more supply if the open surface area of the pond is more. A recording or non-recording raingauge may be set up in the vicinity of the pond to measure the quantity of rainfall falling over the pond. Otherwise the raingauge records of nearby meteorological observatory can be used to compute the magnitude of rainfall. The pond open surface area can be calculated by multiplying the length and width of the pond at ground level for an excavated/dugout pond. Calculations of pond open surface area or sometimes called as top surface area of an embankment type pond is discussed later on.

4.3 Catchment Runoff

Runoff is defined as that portion of precipitation which is not absorbed by the deep soil strata but finds its way into the stream after meeting the demands of evapotranspiration, interception, infiltration, surface storage, surface as well as channel detention. It includes surface, delayed runoff that enters a storage structure after passing through the land surface, and other delayed runoff from the temporarily detained in pits, depressions, swamps or detained as snowfall. It is thus the flow collected from the drainage basin and appearing at the outlet of the basin (Murty, 1998).

4.4 Factors Affecting Runoff

Rate and volume of runoff from an area are influenced mainly by two factors i.e. (i) climatic factors and (ii) physiographic factors. Of the various climatic factors that affect the runoff, nature of precipitation/rainfall is very important.

4.4.1. Climatic Factors

Nature of Precipitation: Surface runoff is produced when the rainfall intensity exceeds the infiltration capacity of the soil. Rainfall intensity influences both the rate and volume of runoff. Even though two storms have same total precipitation, an intense storm of short duration that exceeds the infiltration capacity of a soil at a faster rate produces greater volume of runoff than does a lesser intense storm of longer duration. The intensity-frequency relationship of a storm affects the runoff process in a way that an intense storm occurring frequently produces higher volume of runoff. Similarly a storm of given intensity produces higher amount of runoff if it occurs for a longer duration. In the initial stage of occurrence of a storm, the infiltration capacity may decrease with time and thus a storm of short duration may produce little runoff whereas a storm of longer duration but having same intensity will result in producing larger runoff. Distribution of rainfall also affects the rate and volume of runoff. When there is a higher intense storm occurring uniformly over the whole catchment, the entire watershed contributes and hence, the rate and volume of runoff becomes more. Further if there is an intense storm on one portion of the watershed, then it may produce more runoff than a moderate storm occurring over the entire watershed. If the topography, soil and other conditions in the watershed are uniform, then for any given total volume of rain falling on the watershed, the more non-uniform the distribution is the greater is the peak runoff. The minimum peak is produced by a uniform distribution. Direction of storm movement also affects the peak rate of runoff. A storm moving in the direction of a stream produces a higher peak in a shorter period of runoff than a storm moving upstream. Interception, evaporation and transpiration depend on the meteorological parameters like temperature, humidity wind velocity, and solar radiation and thus influence the runoff. More is the magnitude of these meteorological parameters, more will be losses due to interception, evaporation and transpiration and thus less will be the amount of runoff.

4.4.2. Physiographic Factors

Regarding the size of the catchment, all the factors including depth and intensity of rainfall remaining constant, two catchments irrespective of their sizes will produce the same amount of total runoff expressed as centimeter depth. However, for a catchment with larger area, it takes more time for the runoff to pass through the outlet hence the peak flow expressed a s depth will be smaller. Runoff is influenced by the shape of the catchment. Two catchments having same area, fan shaped compact catchments tend to produce higher peak runoff rate than long and narrow catchments as in the former case, all parts of the catchment contributes to runoff at the outlet simultaneously and thus, the time of concentration to peak runoff is less. On the contrary, fern-leaf shaped catchment gives lesser runoff because the tributaries of different lengths meet the main stream at regular intervals thereby the discharge are likely to be distributed over a large period of time. In addition to the above mentioned two types of catchments, the catchments may be classified depending on the shape as tree like, rectangular type, radial type, parallel type or trellis type.

The slope of the catchment has pronounced affect on infiltration, surface runoff, soil moisture and ground water contribution to stream flow. It affects the period of over land flow. A watershed with flat slope having inadequate drainage systems results in getting more infiltration and hence less runoff than the watershed with steep slope and having well defined drainage systems. Orientation i.e. general direction of the slope of a catchment, with respect to prevailing winds, storm movement, and sun's position has considerable affect on the runoff.

Soil texture and its permeability influences runoff greatly. Shallow pervious soil and coarse textured soil like sandy soil tend to have more infiltration and thereby reducing runoff. Runoff is more for hard compacted subsoil since infiltration is less. Fault zones, fissures and cracks result in enhancing infiltration of in-*situ* rainfall and hence reducing runoff. However, the infiltrated water may contribute to the stream as base flow which is a delayed flow.

Land treatment measures like contour ploughing, strip cropping, bench and terrace cultivation and other soil and water conservation measures increase the water holding capacity of the soil and thus reduces runoff and soil loss in a watershed. Vegetated watersheds, due to, surface retention, and storage produces less runoff than bare watersheds. Storage characteristics like lakes, reservoirs, check dams, pits etc. act as flood modulator and reduce the runoff of a watershed. A channel having rough surface with higher values of roughness coefficient will retard the velocity of flow and thereby reduces the surface runoff contributed to a stream. Topographical factors like altitude and drainage density also affect the runoff. Higher catchments usually have steep slopes and produce greater total runoff. At higher altitudes, though the evaporation rate of the atmosphere is less, the rainfall, under certain condition, is more and hence total runoff may be more. Finally, a watershed with higher drainage density will produce more runoff than a watershed with less drainage density.

4.5 Runoff Estimation

Runoff can be estimated from two sources i.e. due to rainfall and due to snow melt. Various methods used to estimate runoff due to rainfall are:

(i) Runoff coefficient

(ii) Runoff percentage

(iii) Strange's tables and curves

(iv) Empirical formula

(v) Infiltration method

(vi) Rainfall-runoff modelling

4.5.1. Runoff Coefficient

Runoff is a function of rainfall. More is the amount of rainfall, more is the runoff. Thus, the runoff (R) and rainfall (P) can be correlated by a coefficient called as runoff coefficient or runoff factor (K). Runoff coefficient is defined as the ratio of volume of stream flow above the base flow to the volume of rainfall. The rainfall-runoff relation by this approasch is considered as linear as given below:

$$R = K P \qquad \text{... (4.1)}$$

where R is runoff (cm) and P is rainfall (cm). Runoff coefficient (K) depends on all those factors that affect the runoff. Values of K are presented in Table 4.1.

4.5.2. Runoff Percentage

(i) *Binnie's Percentage*: Sir Alexander Binnie has suggested the following percentages of runoff to rainfall based on the data of two rivers of Madhya Pradesh (Table 4.2) (Sharma, 1987).

Table 4.1. Values of runoff coefficient (K) for various soil and vegetation condition

Type of soil/vegetation condition	Values of K	Type of soil/vegetation condition	Values of K
Plateau lightly covered	0.70	Loam largely cultivated and suburbs with garden, lawns	0.30
Clayey soils stiff and clayey soil lightly covered	0.55	Sandy soils, light growth	0.20
Loam lightly cultivated or covered	0.40	Parks, lawn, meadows, gardens	0.05-0.20

Table 4.2. Binnie's percentage formula to convert rainfall to runoff

Annual rainfall (cm)	Runoff (%)	Annual rainfall (cm)	Runoff (%)
50	15	90	35
60	21	100	38
70	25	110	40
80	29		

(i) *Barlow's Percentage*: T.G. Barlow's runoff percentage, based on studies of catchments mostly under 130 km^2 in Uttar Pradesh are given below (Table 4.3).

Table 4.3 Barlow's percentage formula to compute runoff

Class	Description of catchment	Runoff (%)
A	Flat, cultivated and black cotton soils	10
B	Flat, partly cultivated various soils	15
C	Average	20
D	Hills and plains with little cultivation	35
E	Very hilly and steep, with hardly any cultivation	45

The above values of runoff percentages are for average type of monsoon. They are modifies by the application of following coefficients according to the nature of season (Table 4.4).

Table 4.4. Nature of season affecting Barlow's percentage formula to compute runoff

Nature of season	Class of catchment				
	A	B	C	D	E
Light rain	0.70	0.80	0.80	0.80	0.80
Average or varying rainfall with no continuous downpour	1.0	1.0	1.0	1.0	1.0
Continuous downpour	1.50	1.50	1.60	1.70	1.80

4.5.3. Strange's Tables and Curves

Strange, based on the data of Maharastra, has provided following tables and curves between rainfall and runoff. Strange's table (Table 4.5) and curves (Fig. 4.1) give runoff for daily rainfall for good, average and bad catchments and surface conditions being dry, damp and wet prior to rain (Sharma, 1987).

Table 4.5. Daily runoff and percentage of runoff according to Strange for average catchment.

Daily rainfall (mm)	Original stage of ground surface					
	Dry		Damp		Wet	
	Percentage	Runoff (mm)	Percentage	Runoff (mm)	Percentage	Runoff (mm)
5	--	--	4	0.20	7	0.35
10	1	0.10	5	0.50	10	1.0
20	2	0.40	9	1.80	15	3.0
25	3	0.75	11	2.75	18	4.5
30	4	1.20	13	3.90	20	6.0
40	7	2.80	18	7.20	28	11.20
50	10	5.00	22	11.0	34	17.0
60	14	8.46	28	16.8	41	24.6
75	20	15.0	37	27.75	55	41.25
100	30	30.0	50	50.0	70	70.0

N. B. Strange has suggested adding or deducting 25 per cent of runoff values for good or bad catchments.

4.5.4. Rainfall-Runoff Modellling

There are two types of rainfall-runoff modeling i.e. hydraulic methods and lump methods. Hydraulic methods make use of concepts of open channel flow theory which involves in solving the Saint-Venant equations with the appropriate boundary and initial conditions (Chow *et al.*, 1988). The hydraulic methods are tedious and time consuming even using a computer model.

By contrast, the lump methods are much simpler and work surprisingly well especially with irregular small watersheds (Beven, 2001). Two simple lump models have been described in this book; watershed routing technique and synthetic unit hydrograph method.

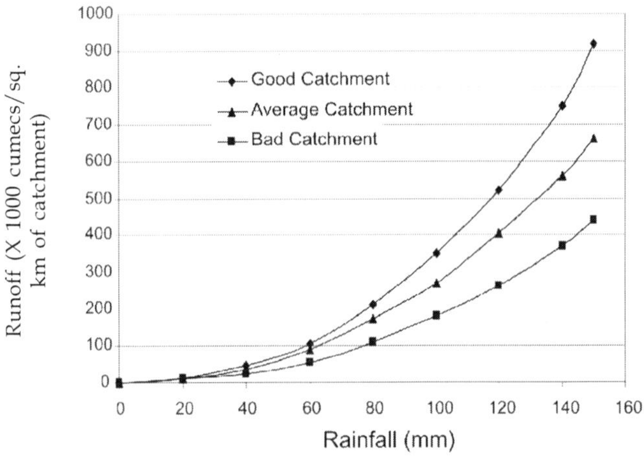

Fig. 4.1. Strange's curves to compute runoff

4.5.4.1. Watershed Routing Technique

The watershed routing model is based on the assumption that the outflow from the watershed varies nonlinearly with the storage in the watershed. An equation as van den Akker and Boomgaard (1996) can be written as:

$$S = KQ^n \qquad \text{... (4.2)}$$

where S = the water stored in the watershed, Q = the outflow from the watershed, and K and n are parameters. By linearizing equation (4.2), the calculation is made much simpler while the accuracy is still acceptable (Beven, 2001), leading to

$$S = KQ \qquad \text{... (4.3)}$$

By converting volume onto depth of water spreading throughout the watershed, we get

$s_1 = kq_1$ ...(4.4)

$s_2 = kq_2$ ...(4,5)

where s_1 and s_2 = depths of water storage at time steps 1 and 2, q_1, q_2 = discharges as depth per unit time at time steps 1 and 2 respectively. From continuity equation,

$s_2 = s_1 + (i - 0.5(q_1 - q_2))\ \Delta t$ ...(4.6)

where i = intensity of excess rainfall, and Δt = time interval. By substituting equations (4.4) and (4.5) into (4.6) and rearranging,

$$q_2 = \frac{k - 0.5\ \Delta\ t}{k + 0.5\ \Delta\ t} \times q_1 + \frac{\Delta\ t}{k + 0.5\ \Delta\ t} \quad ...(4.7)$$

Eqn. (4.7) is the watershed routing model. It can be used to predict the next step flow rate from knowing the present flow rate and intensity of rainfall. The depth flow rate, q, can be converted to volume flow rate, Q, by multiplying q with the watershed area, A.

Watershed routing technique starts with the assumption of the k value. The initial value of depth runoff, q, may be set to zero. Equation (4.7) is used consequently to obtain q of all time steps. The depth runoff value q are converted to the volumetric runoff Q by multiplying q with the watershed area, A. The k value is adjusted until the most suitable hydrograph is approached.

4.5.4.2. Synthetic Unit Hydrograph Method

The unit hydrograph is a direct runoff hydrograph resulting from a unit rainfall (1 cm depth) of specific rainfall duration (Shaw, 1994). The unit hydrograph is normally derived from records of rainfall and runoff data. When dealing with small watersheds, coupled rainfall and runoff data are hardly available. We therefore resort to the synthetic unit hydrograph. The unit hydrograph that is synthesized from topographic and climatic features is called a synthetic unit hydrograph. Essentially, the idea is that the lag time of each watershed is constant and can be evaluated from the watershed characteristics (Shaw, 1994). The lag time is the time lapse between the middle of the rainfall duration and the hydrograph peak. For a small watershed, the lag time can be assumed to be about 0.6 of concentration time.

Time of concentration of a watershed can be computed by Kirpich (1940) formula. Details regarding estimation of runoff from rainfall by synthetic unit hydrograph are reported by Sriboonlue et al. (2003).

In addition to the above mentioned methods, some empirical formulae are also available to estimate the runoff of catchment. Furthermore, infiltration method is also used to estimate the runoff from the record value of rainfall and infiltration loss. Details of estimation of runoff are presented in different text books of Hydrology (Sharma, 1987; Sharma and Sharma, 2002).

4.6 Methods to Estimate Runoff Rate

Following methods are used to estimate runoff rate from a catchment.

(i) Rational Method

(ii) Crook's Method

(iii) Curve Number Method

Out of the above mentioned three methods, the rational and US curve number method are mostly used for computation of runoff rate and hence are discussed in this book.

4.6.1. Rational Method

In this method, the peak rate of runoff is given as:

$$Q = \frac{CIA}{360} \qquad ...(4.8)$$

where, Q is peak runoff rate, m^3/sec; I is rainfall intensity, mm/hr; C is runoff coefficient, and A is catchment area generating runoff, hectares.

Values of runoff coefficient (C) for different land use conditions as determined from the field experiments are given in Table 6.

For catchment with different values of C, the weighted value of C as estimated below is considered for the whole catchment.

$$C = \frac{\sum_{i=1}^{n} C_i A_i}{A} \qquad ...(4.9)$$

where, C is the runoff coefficient of the whole catchment, A is the area of the whole catchment, C_2 is the runoff coefficient for catchment i and Ai is the area of catchment i for which runoff coefficient is C_2, and i is the index for catchment ranging from number 1 to n.

Table 4.6. Values of runoff coefficient, C used in rational method

Soil Types	Land Use		
	Cultivation	**Pasture**	**Forest**
Sandy or gravelly soil having high infiltration rate	0.29	0.15	0.10
Loamy soil and soils with no clay pans near the surface having average infiltration rate	0.40	0.35	0.30
Heavy clay soil or soils with clay pan near the surface or shallow soil above impervious rock having below average infiltration rate	0.50	0.45	0.40

The value of the intensity of rainfall to be used in Eq. 4.8 should be calculated for the period equal to the time of concentration of the catchment. While using the rainfall intensity in the rational method, the rainfall records at different places are analysed and the 1 hour rainfall intensity are calculated. Since small water harvesting structures including farm ponds are generally designed for 10 years return periods, the rainfall intensity at 10 years frequency is considered for computation of runoff in rational formula. The values of maximum rainfall at 1 hour duration at 10 years frequency is given in Fig. 4.2. This 1 hour intensity is converted to the intensity for the duration of time of concentration by using the graphs as proposed by Tejwani et al. (1975) and is given in Fig. 3. The time of concentration of the catchment is defined as the time required for water to flow from the remotest point of the catchment to outlet of the catchment.

When the duration of the rainfall equals to the time of concentration, all parts of the catchment will take part in contributing to the runoff and hence the discharge would be maximum.

Kirpich (1940) has given the following empirical formula to compute the time of concentration.

$$T_c = 0.02\ L^{0.77}\ S^{-0.385} \qquad ...(4.10)$$

where, Tc is time of concentration, min; L is the length of the channel reach, m and S is the average slope of the channel reach, m/m.

For small catchment with area less than 5 sq. km having less developed network of channels and where overland flow dominates, the time of concentration is given by the Eq. 4.11 (Haan *et al.*, 1982) as:

$$T_c = 0.02\, L^{0.77}\, S^{-0.385} + \left(\frac{2L_o n}{\sqrt{S_o}}\right)^{0.467} \qquad ...(4.11)$$

where, Lo is the length of the overland flow, m; So is the slope along the flow path, m/m and n is the average roughness coefficient of the channels contributing overland flow.

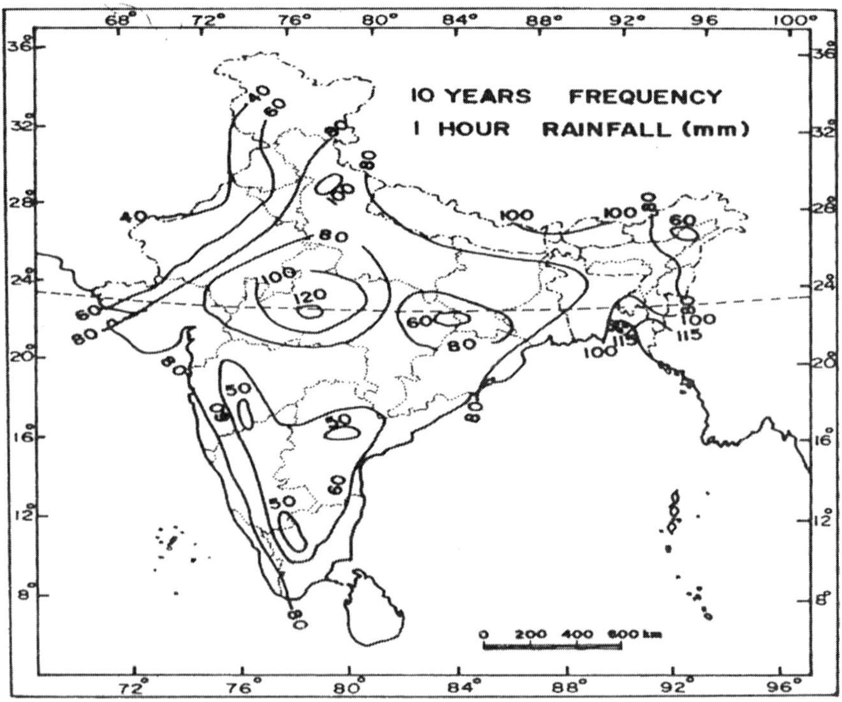

Fig. 4.2. One hour maximum rainfall at 10 years frequency

The second term of Eq. 4.11 accounts for the overland flow. The value of this term has a maximum value regardless of watershed size. The value of n suggested by Haan *et al.* (1982) is given in Table 4.7.

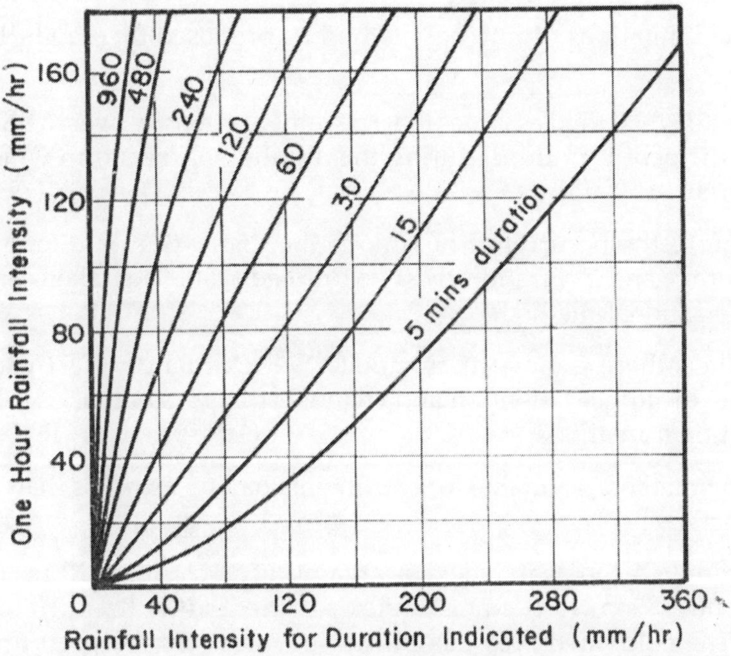

Fig. 4.3 Curves to convert one hour rainfall intensity to intensity of other duration

Table 4.7. Values of roughness coefficient n used in overland flow

Nature of the surface	Values of n
Smooth, impervious surface	0.02
Smooth, bare packed surface	0.10
Cultivated row crops of moderately rough bare soil	0.20
Pasture or average grass	0.40
Deciduous forest areas	0.60
Timber land with deep forest litter or dense grass	0.80

Time of concentration can also be computed as:

$$Tp = Tc^{0.5} + 0.6\ T_c \qquad ...(4.12)$$

where, Tp is time from beginning of rise of hydrograph to the peak.

Limitations of Rational Method

The limitations of rational method as proposed by (Singh, 1964) are:

(i) Rational method assumes that the rainfall intensity over the whole cathcment is uniform during the duration of the storm which is rarely satisfied.

(ii) Before the occurrence of runoff, the losses due to depression storage and infiltration must be fulfilled that is not taken care by the rational method.

(iii) When there is flow in channels, the retardance affect caused due to surface detention and channel storage is not taken care in rational method.

(iv) The method is suitable for catchments having area less than 5000 hectares.

Example 4.1 A watershed of area 50 hectares have 20 hectares under cultivation, 10 hectares under pasture and 20 hectares under forest. The cultivated area has sandy soil, whereas the pasture and forest area have heavy clay soil with very low infiltration rate. The watershed has a fall of 10 m over the longest reach of length 1 km; the reach being connecting the remotest point to the outlet point of the watershed. Estimate the peak runoff rate if the 1 hour rainfall intensity for 10 years frequency of the watershed is 100 mm/hr. Assume that the ratio of 1 hour rainfall intensity to rainfall intensity of any other duration is 1.2,

Solution: Using Table 4.6, the values of C for cultivated, pasture and forest areas are calculated as 0.29, 0.45 and 0.40, respectively. Weighted value of C for the whole watershed is calculated as

$$C = \frac{20 \times 0.29 + 10 \times 0.45 + 20 \times 0.40}{50} = 0.366$$

For L = 1 km = 1000 m, slope S = 10/1000 = 0.01

Using Eq. 4.10, time of concentration, T_c = 24 min.

Given the one hour rainfall intensity of 10 year frequency is 100 mm/hr and the ratio of 1 hour rainfall intensity to that of any other duration is 1.2, the rainfall intensity for 24 minutes of rainfall is 100 x 1.2 = 120 mm/hr.

$$\text{Hencepeak runoff rate} = = Q = \frac{CIA}{360} \quad \frac{0.366 \times 120 \times 50}{360} = 6.1\,\text{cumecs}$$

4.6.2. Curve Number Method

This method is also called as Hydrologic Soil Cover Complex Number method or US Soil Conservation Service (SCS) method. It is based on the recharge capacity of the watershed. The recharge capacity is determined by the antecedent moisture condition (AMC) and by the physical characteristics of the watershed. Let I_a be the initial quantity of interception, depression storage and infiltration that must be satisfied by any rainfall before runoff can occur. It is assumed that the ratio of the direct runoff Q and the rainfall minus initial losses (P Ia) and the storage capacity Sr are related by:

$$\frac{Q}{P - I_a} = \frac{P - Q - I_a}{Sr} \qquad ...(4.13)$$

I_a is assumed to be certain percentage of Sr. Assuming it to be 0.2 Sr, the runoff is given as

$$Q = \frac{(P - 0.2Sr)^2}{(P + 0.8Sr)} \quad \text{if } P > 0.2Sr \qquad ...(4.14)$$

$$Q = 0 \quad \text{if } P \leq 0.2Sr \qquad ...(4.15)$$

where, Q = surface runoff, mm; P = rainfall, mm and Sr = potential maximum retention, mm.

The parameter Sr (value in mm) is related to the curve number (CN) as:

$$Sr = 254\left(\frac{100}{CN} - 1\right) \qquad ...(4.16)$$

The curve number for antecedent moisture condition (CN for AMC II or CN = CN2) for different land use conditions and hydrologic soil groups are given by USDA (1972) (Table 4.8). Fig.4.4 helps in the computation of runoff from rainfall for different values of curve number (CN) for AMC II. A heterogeneous basin may be divided into sub areas with different values of curve numbers and in such case, a weighted value of CN for AMC II is calculated for the whole watershed. The corresponding values of CN for dry (CN for AMC I = CN1) and wet (CN for AMC III = CN3) conditions are given by Sharpley and Williams (1990) as:

$$CN_1 = CN_2 - \frac{20(100 - CN_2)}{100 - CN_2 + \exp\{2.533 - 0.0636(100 - CN_2)\}} \quad ...(4.17)$$

$$CN_3 = CN_2 + 10\exp\{0.00673(100 - CN_2)\} \quad ...(4.18)$$

and Sr given by them is:

$$Sr = Sr_1\left[1 - \frac{FFC}{FFC + \exp(W_1 - W_2.FFC)}\right] \quad ...(4.19)$$

where, Sr_1 is the value of Sr associated with CN1 and FFC is the availability of soil water expressed as fraction of field capacity given as:

$$FFC = \frac{SMC_i - WP}{FC - WP} \quad ...(4.20)$$

where SMC_2 is the soil moisture content on day i, FC and WP are the soil moisture content at field capacity and wilting point, respectively.

W_1 and W_2 are given as:

$$W_1 = \ln\frac{Sr_3}{Sr_1 - Sr_3} + W_2 \quad ...(4.21)$$

$$W_2 = \ln\left(\frac{0.5Sr_2(Sr_1 - Sr_3)}{Sr_3(Sr_1 - Sr_2)}\right)^2 \quad ...(4.22)$$

Table 4.8. Curve numbers for hydrologic soil cover complexes

(For watershed condition II I_a=0.2S) Land Use or cover	Treatment or Practice	Hydrologic Condition	Hydrologic Soil Group			
			A	B	C	D
Fallow- Row	Straight row		77	86	91	94
Crops	Straight row	Poor	72	81	88	91
	Straight row	Good	67	78	85	89
	Contoured	Poor	70	79	84	88
	Contoured	Good	65	75	82	86
	Contour & terraced	Poor	66	74	80	82
	Contour & terraced	Good	62	71	78	81
Small Grain	Straight Row	Poor	65	76	84	88
	Straight	Good	63	75	83	87
	Contour	Poor	63	74	82	85
	Contour	Good	63	75	83	87
	Contoure & terraced	Poor	61	72	79	82
	Contour & terraced	Good	59	70	78	81
Seeded Legumes	Straight row	Poor	66	77	85	89
or Rotation	Straight row	Good	58	72	81	85
Meadow	Contour	Poor	64	75	83	85
	Contour	Good	55	69	78	83
	Contour & terraced	Poor	63	73	80	83
	Contour & terraced	Good	61	67	76	80
Pasture or		Poor	68	79	86	89
range		Fair	49	69	79	84
		Good	39	61	74	80
	Contour	Poor	47	67	81	88

Contd...

Table 4.8 Contd...

	Contour	Fair	25	59	75	83
	Contour	Good	06	35	70	79
Meadow (permanent)		Good	30	58	71	78
Woodlands		Poor	45	66	77	83
(farm woodlots)		Fair	36	60	73	79
Farmsteads			59	74	82	86
Roads(dirt)			72	82	87	89
Roads(Hard surface)			74	84	90	92

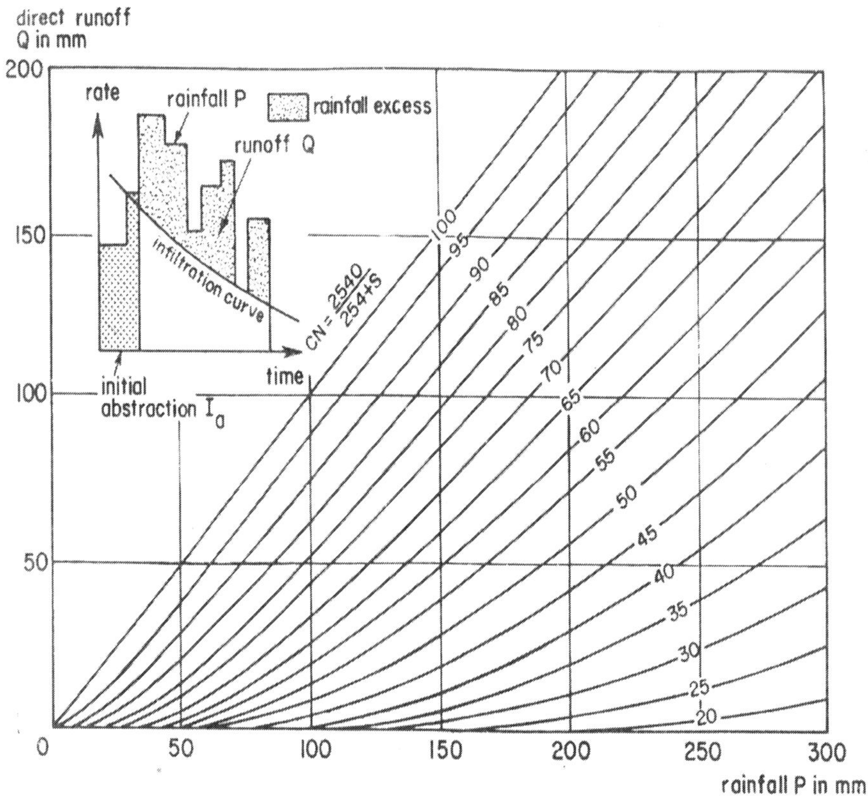

Fig. 4.4. Curve number to convert rainfall to runoff

In which Sr_2 and Sr_3 are the values of Sr when FFC = 0.5 and 1, respectively. Sr_1, Sr_2 and Sr_3 are the retention parameters corresponding to CN1, CN2 and CN3, respectively. Once the parameters W_1, W_2 and FFC are estimated, value of Sr can be obtained from Eqn. 4.19. The estimates of peak runoff rate from total runoff computed by Eq. 4..14 as suggested by US Soil Conservation Service (1972) is:

$$Q_{peak} = \frac{0.0208AQ}{T_p} \quad ...(4.23)$$

where, Q_{peak} is peak rate of runoff, cumecs; A is area, hectares; T_p is time to peak, hours; T_c is time of concentration, hours discussed earlier in Eqs. 4.10 and 4.11 and Q is runoff, cm. T_p is given as

$$T_p = \sqrt{T_c} + 0.6T_c \quad ...(4.24)$$

Antecedent moisture condition (AMC) is an index for watershed wetness. The three levels of AMC are:

AMC I: Lowest runoff potential with dry condition of watershed.

AMC II: Average condition.

AMC III: Highest runoff potential with wet condition of watershed.

The AMC class is determined using 5 days antecedent rainfall in dormant or crop growing season as given below (Table 4.9).

Table 4.9. Values of 5 days antecedent rainfall in deciding AMC class

AMC	5 days antecedent rainfall, cm	
	Dormant season	**Growing season**
I	< 1.25	< 3.5
II	1.25 - 2.75	3.5 - 5.25
III	> 2.75	> 5.25

The hydrologic soil groups are divided into 4 classes depending on runoff potential as :

Hydrologic Soil Group A (Low runoff potential): Soil having high infiltration rate and consisting chiefly of deep well to excessively drained sands or gravels.

Hydrologic Soil group B (Moderately low runoff potential): Soil having moderate infiltration rate and consisting chiefly of moderately deep to deep, moderately well to well drained soils having moderate coarse texture.

Hydrologic Soil group C (Moderately high runoff potential): Soil having low infiltration rate and consisting chiefly of moderately deep to deep, moderately well to well drained soils having moderate coarse texture.

Hydrologic Soil group D (High runoff potential): Soil having very low infiltration rate and consisting chiefly of clay soils with high swelling potential and soil with clay layer at or near the surface. Hydrologic soil groups as mentioned above is given by Gupta et al. (1970) and is given in Table 4.10.

Example 4.2: A watershed of area 50 hectares consist of 20 hectare with row crops cultivated in good terrace land and 30 hectares of good pasture land. The soil of the watershed belongs to hydrologic soil group C. Values of field capacity and wilting point soil moisture contents are 20 and 10%, respectively and soil moisture content of the soil on a particular day is 16%. Find the runoff amount on that day when a rainfall amount of 100 mm occurs in the watershed.

Solution: Using the data of land use, treatment conditions, hydrologic soil group and hydrologic condition, values of CN (AMC II) for 20 and 30 hectares lands are found from Table 4.8 as 78 and 74, respectively. The weighted CN value is given as

$CN_2 = (78 \times 20 + 74 \times 30)/50 = 75.6$

Using Eq. 4.17, $CN_1 = 57.68$ and Eq. 4.18, $CN_3 = 87.40$

Value of Sr_1 for $CN_1 = 57.68$ using Eq. 4.16 (replacing CN as CN_1) = 186.36. Similarly using Eq. 4.16 and replacing CN as CN_3, value of Sr_3 is obtained as 36.62.

Value of FFC (Eq. 4.20) = (16-10)/(20-10) = 0.6

Using Eq. 4.22, W2 is obtained as 0.944 and using Eq. 4.21, W1 is obtained as -0.461.

Using Eq. 4.19, value of Sr is obtained as 69.63 mm

Using Eq. 4.14, Q is obtained as

$$Q = \frac{(100 - 0.2 \times 69.63)^2}{100 + 0.8 \times 69.63} = 47.58 \text{ mm.}$$

Table 4.10. Hydrological grouping of different soil types

S. No.	Soil type	Tentative hydrological groups
1	Alluvial soil non calcareous to moderately calcareous	B
2	Alluvial soil highly calcareous	B
3	Alluvial soil developed on coast as alluvium	A
4	Alluvial soil developed on deltaic alluvium	B
5	Alluvial soil reverine affected by salinity and alkalinity	C
6	Pedocal sicrozem of alluvial origin	C
7	Pedocal brown soil of alluvial origin	C
8	Grey and brown (desert soil)	D
9	Desert soil	A
10	Deep black soil	D
11	Medium black soil	D
12	Shallow black soil	D
13	Black soil affected by salinity and alkalinity	D
14	Black soil undifferentiated	D
15	Mixed red and black soil	C
16	Ferruginous red soil	B
17	Ferruginous red gravely soil	B
18	Red and yellow soil	C
19	Laterite	B
20	Laterite and black soil	D
21	Brown soil under deciduous forest	B

Contd...

Table 10.Contd...

22	Forest soil	B
23	Podsolic soil	B
24	Foothill soil	C
25	Mountain and hill soil	B
26	Mountain meadow soil	B
27	Peat	B

Question Banks

Q1. What are the sources of inflow to a farm pond?

Q2. How is the *in situ* rainwater contribution to a farm pond decided?

Q3. What do you mean by runoff? What are the factors on which the runoff of catchments depends on?

Q4. How is the nature of precipitation affect the runoff of a catchment?

Q5. Explain the different physiographic factors that influence the runoff of a catchment?

Q6. Which one of the following will produce more runoff and explain why?

(i)Fan shaped catchment and (ii) Fern shaped catchment.

Q7. What do you mean by rainfall-runoff modeling? Explain the watershed routing method to estimate the runoff from a given rainfall mount.

Q8. Briefly explain the different methods used to compute the runoff produced from a catchment.

Q9. Describe the rational method to estimate the runoff rate of a catchment.

Q10. What are the limitations to use the rational method for computation of runoff rate of a catchment?

Q11. A watershed has an area of 100 ha under pasture, 120 ha under cultivation and 50 ha under the forest. The cultivated area has loamy soil whereas the pasture and forest lands have heavy clay soil with very low infiltration rate. The average longitudinal slope of the watershed with the longest reach contributing runoff is0.1 percent. The one hour rainfall intensity for 10 years frequency interval of the watershed is 60 mm/hr. Estimate the runoff rate of the watershed. Assume the ratio of one hour rainfall intensity to rainfall intensity of any other duration as 1.3.

Q12. Explain how the runoff rate of a catchment is estimated by the curve number method.

Q13. Write about the four hydrologic soil groups. What are their characteristic features?

Q14. Which one of the following will produce more runoff rate and explain why?

(i)Hydrologic soil group A and (ii) Hydrologic soil group B.

Q15. A watershed has an area of 100 ha out of which 60 ha is under the row crop with good terrace and the rest is under good pasture land. Soil of the watershed belongs to the hydrologic soil group B. Values of field capacity and wilting point soil moisture contents are 22 and 10%, respectively and soil moisture content of the soil on a particular day is 18%. Find the runoff amount on that day when a rainfall amount of 120 mm occurs in the watershed.

Q16. What do you mean by the term AMC? What are the factors on which the AMC depends on?

Q17. What do you mean by the term CN? Write in short the factors on which the CN depends on?

Q18. What is lag time and how is it determined?

References

Beven, K.J. 2001. Rainfall-Runoff Modelling: *The Primer*. Wiley.

Chow, V.T.; Maidment, D.R. and Mays, L.W. 1988. Applied Hydrology. McGraw-Hill.

Gupta, S.K., Ram Babu and Tejwani, K.G. 1970. Nomograph and important parameters for estimation of peak rate of runoff from small watershed. Indian Journal of Agricultural Engineering, ISAE, New Delhi, Vol. 8(3): 25-32.

Hann, C.T., Johnson, H.P. and Brakensick, D.L. 9Editors) 1982. Hydrologic Modelling of Small Watersheds. American Society of Agricultuural Engineers, St. Joseph.

Kirpich, Z.P. 1940. Time of concentration of small agricultural watersheds. Civil Engg., 10:362.

Murty, V.V.N. 1998. Land and Water Management Engineering. 2nd Edition, Kalyani Publishers, New Delhi, pp. 586.

Sharma, R.K. and Sharma, T.K. 2002. Irrigation Engineering including Hydrology, S. Chand and Company Ltd. New Delhi.

Sharma, R.K., 1987. A Text book of Hydrology and Water Resources. Dhanpat Rai and Sons, Delhi.

Sharpley, A.N. and Williams, J.R. (Edn.) 1990. EPIC-Erosion/Productivity Impact Calculator: (1) Model Documentation, pp.235.

Shaw, E.M. 1994. Hydrology in Practice. 3rd ed., Chapman and Hall.

Singh, J. 1964. Irrational use of the Rational formula. Er presented at the third annual meeting of the Indian Society of Agricultural Engineers, Allahabad.

Sriboonlue, V., Khetkratok, N and Polkern, K. 2003. Inflow calculation for on-farm ponds in northeast Thailand, Comprehensive Assessment of Water Management in Agriculture Series, Water Productivity in Agriculture: Limits and Opportunities for Improvement. Book Edited by J.W.Kijne, R. Barker and D. Molden. CABI Publishing, pp. 352.

Tejwani, K.G., Gupta, S. K. and Mathur, H.N. 1975. Soil and Water Conservation Research 1956-1971. Indian Council of Agricultural Research, New Delhi.

US Soil Conservation Service.1972. National Engineering Handbook, Section 4, Hydrology, Department of Agriculture, Washington, D.C.

van den Akker, C. and Boomgaard, M.E. 1996. Hydrologie. Technische Universiteit Delft.

Chapter 5

Water Demand From Farm Pond

5.1 Introduction

A farm pond that collects and harvests *in-situ* rainfall as well as runoff from the catchment area can be used for many purposes: the most important of which is the supply of water to crops as supplemental irrigation. It is no doubt a viable source for integrated farming system. In addition, the water of the pond can be used for domestic and to some extent for the industrial uses. It is also a good source for recreational use. Total water demand of the pond including irrigation to different crops in different seasons, domestic and other non agricultural uses are to be assessed. While estimating the crop water requirement and irrigation requirement, various crops to be grown in the service area of the pond, their water requirements, crop duration, extent of coverage and water management practices followed are to be taken into consideration. Multiplying the irrigation requirement with crop coverage area, volumetric irrigation water requirement of crop can be found out. Summation of irrigation requirement of all crops grown in service area of the pond, gives total irrigation demand of the pond.

Farm ponds are normally used for irrigation to crops as well as to meet the live stock water demand. Normally the large live stocks like horse, cattle require 20 thousand litres of water per head per year whereas the small live stocks like sheep, goat, pigs require 2 thousand litres of water per head per year. Adding the irrigation demand to the livestock demand, total water requirement of the pond is calculated. In addition to this, if the pond water is used for industrial, domestic and other sectoral purposes, then water used in all such sectors are to be added to irrigation and livestock demand to estimate the total water demand of the pond.

5.2 Water and Irrigation Demand of Crops

Of the various demand of water from the pond, irrigation demand to crops grown in the command area of the pond is very important. Most of the crop water demand is met from rainfall. However, when the rainfall is deficient to meet the crop water demand, then the deficit is met from the stored water of the pond in the form of irrigation, more commonly called as supplemental irrigation. The irrigation to the crops is required mostly to meet the evapotranspiration requirement of the crop. Before discussing the water and irrigation demand/ requirement of crops, it is essential to know some commonly used terms frequently used. They are:

Consumptive Use: It is the quantity of water lost in evaporation from the soil and water surface, transpiration by the vegetations grown on the soil and that used in metabolic activity. Water used in metabolic activity is very less (usually less than 1% of evapotranspiration). Thus, in usual sense, the term consumptive use is taken as evapotranspiration, ET.

Water Requirement: It is the water needed for raising a crop in a given period starting from seedling to harvest. It includes consumptive use and other unavoidable losses and also the water required for special operations like land preparation, transplanting and leaching.

Evapotranspiration: It is the total amount of water lost by the evaporation from the soil and water surface and transpiration by the vegetations. It depends on the crop types, its coverage amount and management and cultural practice of the crop.

Irrigation Requirement: It refers to the quantity of water required to supply for successful growth of plants exclusive of precipitation, groundwater contribution and other natural resources.

Effective Rainfall: It is that part of rainfall which forms part of water requirement. Effective rainfall depends on a number of factors. It depends on the amount, intensity and distribution of rainfall event, antecedent soil moisture content, crop cover and management practice, and the season. A light intensity rainfall in winter may be 100% effective whereas a high intensity rainfall in rainy season occurring for a small duration may be 30-50% effective.

Duty of Water: Relation between irrigation water and area of crop that matures fully with amount of water is called as duty of water. It is usually expressed as area per unit rate of flow.

Delta: It refers to the depth of water needed for raising the crops excluding rainfall for full crop period.

5.3 Methods to Estimate Crop Water Requirement

The most important component of water requirement of crop is its evapotranspiration. There are several methods used to estimate the crop evapotranspiration. Broadly they can be classified as:

(i) Estimation by energy balance method

(ii) Direct measurements and

(iii) Estimation by climatologically approach.

5.3.1. Estimation by Energy Balance Method

Based on the principles of conservation of energy and considering a soil column extending from the surface to some depth where vertical heat exchange is negligible, the energy balance can be written (Michael, 1987) as

$$(Q + q)\ (1 - a) + I\downarrow - I\uparrow + H + LE + G + M = 0 \qquad ...(5.1)$$

where, Q = direct solar radiation incident on earth; q = diffuse solar radiation incident on the earth; a = surface resistance (albedo) which is a fraction; $I\downarrow$ = incoming long wave radiation; $I\uparrow$ = outgoing long wave radiation from the land; H = sensible heat transfer; LE = latent heat flux due to evaporation; G = net rate at which heat content of soil column is changing and M = miscellaneous heat terms.

Eq. (5.1) can be used for any time period. The units are generally expressed in cal/cm^2/unit time. It is to be noted that 1 cal/cm^2/min = 1 mm of water evaporating per hour. In Eq. (5.1) Rn = net solar radiation which is the difference between the total incoming and outgoing solar radiation.

Neglecting M, Eq. (5.1) can be written as:

$$R_n = H + LE + G \quad \text{...(5.2)}$$

5.3.2. Direct Measurement of Evapotranspiration

(i) Lysimeter experiment

(ii) Field experimental plots

(iii) Soil moisture depletion studies and

(iv) Water balance method.

5.3.2.1. Lysimeter Method

Lysimeter studies involve the growing of crop in large containers (lysimeters) and measuring their water loss and gains. Mainly there are two types of lysimeters (Fig. 5.1). They are (i) weighing type and (ii) drainage type. Weighing type lysismetrs give more accurate results compared to the drainage type. To get more accurate and short period results, weighing type lysimeter can be used. The soil and crop conditions of the lysimeters should be close to the natural condition. From the irrigation point of view, weighing type lysimeter is set up to enable the operator to measure the water balances: water added, water retained by the soil, and water lost through all sources-evaporation, transpiration and deep percolation. These measurements require weighing that may be made with scales or by floating the lysimeters in water in which case the change of liquid displacement is computed against water loss from the tank. The techniques yield measurement of total water loss and is useful as an indicator of field water loss. The precautions needed in this case is that the tank must be permanently buried in the ground and surrounded by a large area of crop with same height as crop. The water table is maintained as specific depth in the tank by connecting it to a supply tank provided with a float mechanism which has an arrangement for receiving excess water that tends to build up in the tank. Water is applied in measured amount to the lysimeter as irrigation is applied to the crops in the fields where lysimeter is set up. The water received either from the reservoir or precipitation excluding the outflow, constitutes the water used by the crops (Lenka, 2001).

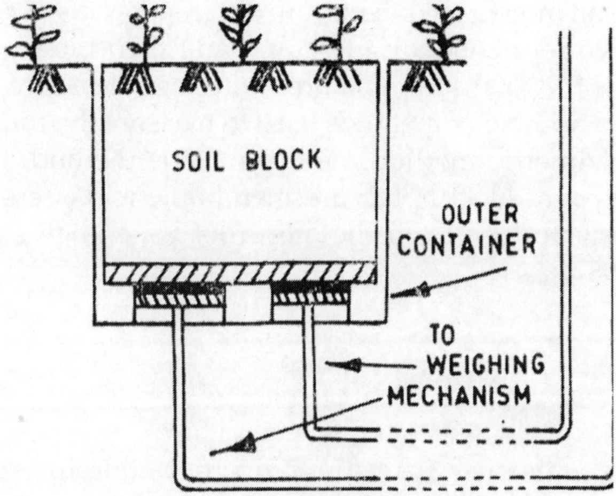

(a) Weighing type lysimeter

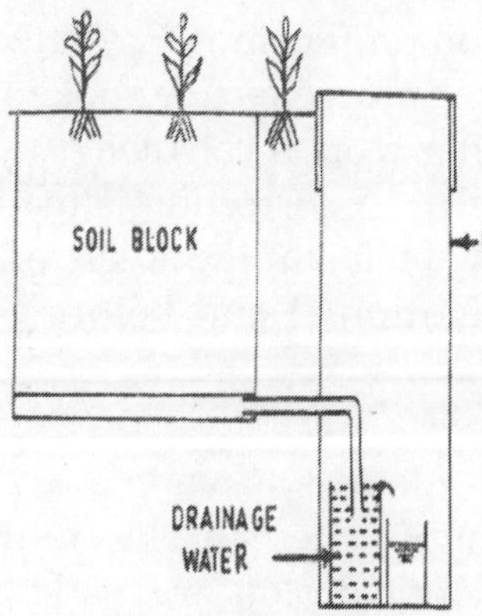

(b) Drainage type lysimeter

Fig. 5.1. Lysimeters to measure water requirement of crops

5.3.2.2. Soil Moisture Depletion Studies

This method is usually suitable for irrigated areas where soil is fairly uniform and the depth to the groundwater is such that it will not influence the soil moisture fluctuations within the crop root zone. In this method, the soil moisture within the crop effective root zone before and just after irrigation and also within each irrigation cycle is measured. The soil moisture content from different profile layers within the crop effective root zone are measured and then these values are used to compute the total soil moisture content in the entire crop root zone. There are a number of methods used to measure the soil moisture content i.e. gravimetric method, neutron probe method, infra red, tensiometer, gypsum block, pressure membrane and pressure plate technique. The water use from the crop root zone is given (Mishra and Ahmed, 1987) as:

$$u = \sum_{i=1}^{n} \frac{(M_{1i} - M_{2i})}{100} .A_i.D_i + ER - Dp \qquad ...(5.3)$$

where, u = water use from the crop root zone for successive sampling periods or within one irrigation cycle, mm; n = number of soil profile layers in the crop effective root zone D; M_{1i} = soil moisture percentage at the time of the first sampling in the ith layer; M_{2i} = soil moisture percentage at the time of the second sampling in the ith layer; A_i = apparent specific gravity of the ith layer of the soil and D_i = depth of the ith layer of the soil profile, mm, E_R = effective rainfall, mm and Dp = percolation from the crop effective root zone, mm. The seasonal value of water use by the crop is calculated by adding the water use of each sampling interval. A correction is made by adding the potential evapotranspiration value(s) for the accelerated water loss for the interval(s) just after irrigation(s) and before soil moisture sampling.

5.3.2.3. Field Experimental Plots

In this method, measurement of water supplies to the fields and change in soil moisture contents of field plots are made. The seasonal water requirement is computed by adding measured quantities of irrigation water, effective rainfall received during the season and contribution of soil moisture form the crop effective root zone. This is expressed by the relationship (Mishra and Ahmed, 1987):

$$WR = IR + ER + \sum_{i=1}^{n} \frac{(M_{bi} - M_{ei})}{100} .A_i .D_i \qquad ...(5.4)$$

where, WR = seasonal water requirement, mm; IR = total irrigation water supplied, mm; ER = seasonal effective rainfall, mm; M_{bi} and M_{ei} = soil moisture percentage at the beginning and end of the season in the ith layer of the soil, respectively and other terms of equation are as discussed above. This method has the limitation that it does not provide information on intermediate soil moisture conditions, short term use, profile use, deep percolation and peak use rate of the crop.

5.3.2.4. Water Balance Method

The method is suitable for large fields over long periods. It is based on inflow and outflow of water to the soil control volume. The water balance model of a crop field is expressed (Panigrahi, 2001) as:

$$S_{w,i} = S_{w,i-1} + R_i + I_{s,i} - E_{Ta,i} - Q_i - S_i - D \qquad ...(5.5)$$

where, R = rainfall; I_s = supplemental irrigation, E_{Ta} = evapotranspirationof the crop; Q = runoff from the crop field; D = percolation loss in the crop field, S = seepage loss from the crop filed and S_{wi} and S_{wi-1} = soil moisture content on ith and (i-1)th day, respectively. The term i of the above equation refers to the day which may be taken as 1 day. The units of various terms in the above equation is mm. It is to be remembered that the capillary rise due to groundwater table is ignored (Odhiambo and Murty, 1996).

Measuring all the parameters of the water balance model (Eq. 5.5) except crop evapotranpiration, we can calculate E_{Ta}. The method has the advantage that we can compute the crop evapotranspiration on short term basis like daily basis.

5.3.3. Estimation by Climatologically Approach

Owing to the difficulty in measuring the different water balance parameters including the data required for computation of crop evapotranspiration by lysimeters, estimation methods by busing climatic data are preferred. There are several climatologically factors which will influence and decide the rate of evaporation. In fact when water is not a limiting factor, the rate of evaporation is decided by atmospheric demand of the water. Some of the important factors of climate influencing the evaporation are radiation, temperature, humidity, and wind speed. The estimation methods are empirical methods. Some of the estimation methods require only one or two climatic data and does not give accurate result of evapotranspiration. However, there are other estimation methods like FAO Penmain-Monteith model which require a lot of climatic data and give accurate result of evapotranspiration. The empirical method/estimation method involves estimating the reference crop evapotranspiration and then multiplying the reference crop evapotranspiration with crop coefficient we can estimate the crop evapotranspiration. Some of the more commonly used empirical methods to estimate reference crop evapotranspiration are given below. For detail estimation of various empirical methods, readers are suggested to refer FAO books (Doorenbos and Pruitt, 1977; Allen et al., 1998).

5.3.3.1. Thornwaite Method

Thornwaite (1948) assumed that the amount of water lost through evapo-transpiration from soil surface covered with vegetation is governed by climatic factors and is independent of species when moisture supply is not limiting. He proposed the following formula:

$$e = 1.6\left(\frac{10\ t}{I}\right)^{a} \qquad ...(5.6)$$

where, e = unadjusted potential evapotranspiration, cm/month

t = mean air temperature, °C

I = annual or seasonal heat index, the summation of 12 values of monthly heat indices (i) where I = $\left(\frac{t}{5}\right)^{1.514}$ and a = an empirical exponent computed by the equation

$$a = 0.000000675\ I^3 - 0.0000771\ I^2 + 0.01792\ I + 0.49239$$

The factor e is an unadjusted value based on a 12 hour day and 30 day month and is corrected by actual day length in hours and day in a month, to give the adjusted potential evapotranspiration PET (reference evapotranspiration) as.

PET = e x correction factor

The correction factor for different latitudes and months are provided in Table 5.1.

Table 5.1. Correction factors to be used in Thotnthwaite formula (Mean possible duration of sunlight expressed in units of 30 days of 12 hours each)

North Latitude	Jan	Feb	Mar	Apr	May	June	Jul	Aug	Sept	Oct	Nov	Dec
0	1.04	0.94	1.04	1.01	1.04	1.04	1.04	1.04	1.01	1.04	1.01	1.04
5	1.02	0.93	1.03	1.02	1.06	1.03	1.06	1.05	1.01	1.03	0.99	1.02
10	1.00	0.91	1.03	1.03	1.08	1.06	1.08	1.07	1.02	1.02	0.98	0.99
15	0.97	0.91	1.03	1.04	1.11	1.08	1.12	1.08	1.02	1.01	0.95	0.97
20	0.95	0.9	1.03	1.05	1.13	1.11	1.14	1.11	1.02	1.00	0.93	0.94
25	0.93	0.89	1.03	1.06	1.15	1.14	1.17	1.12	1.02	0.991	0.9	0.91
26	0.92	0.88	1.03	1.06	1.15	1.15	1.17	1.12	1.02	0.99	0.91	0.91
27	0.92	0.88	1.03	1.07	1.16	1.15	1.18	1.13	1.02	0.99	0.90	0.90
28	0.91	0.88	1.03	1.07	1.16	1.16	1.18	1.13	1.02	0.98	0.90	0.90
29	0.91	0.87	1.03	1.07	1.17	1.16	1.19	1.13	1.03	0.98	0.90	0.89
30	0.9	0.87	1.03	1.08	1.18	1.17	1.20	1.14	1.03	0.98	0.89	0.88
31	0.9	0.87	1.03	1.08	1.18	1.18	1.20	1.14	1.03	0.98	0.89	0.88
32	0.89	0.86	1.03	1.08	1.19	1.19	1.21	1.15	1.03	0.98	0.88	0.87
33	0.88	0.86	1.03	1.09	1.19	1.20	1.22	1.15	1.03	0.97	0.88	0.86
34	0.88	0.85	1.03	1.09	1.20	1.20	1.22	1.16	1.03	0.97	0.87	0.86
35	0.87	0.85	1.03	1.09	1.21	1.21	1.23	1.16	1.03	0.97	0.86	0.85
36	0.87	0.85	1.03	1.10	1.21	1.22	1.24	1.16	1.03	0.97	0.86	0.84
37	0.86	0.84	1.03	1.10	1.22	1.23	1.25	1.17	1.03	0.97	0.85	0.83
38	0.85	0.84	1.03	1.10	1.23	1.24	1.25	1.17	1.04	0.96	0.84	0.83
39	0.85	0.84	1.03	1.11	1.23	1.24	1.26	1.10	1.04	0.96	0.84	0.82
40	0.84	0.83	1.03	1.11	1.24	1.25	1.27	1.18	1.04	0.96	0.83	0.81
41	0.83	0.83	1.03	1.11	1.25	1.26	1.27	1.19	1.04	0.96	0.82	0.80
42	0.82	0.83	1.03	1.12	1.26	1.27	1.28	1.19	1.04	0.95	0.82	0.79
43	0.81	0.82	1.02	1.12	1.26	1.28	1.29	1.20	1.04	0.95	0.81	0.77
44	0.81	0.82	1.02	1.13	1.27	1.29	1.30	1.20	1.04	0.95	0.80	0.76

Cond...

Table 5.1.Cond...

45	0.80	0.81	1.02	1.13	1.28	1.29	1.31	1.21	1.04	0.94	0.79	0.75
46	0.79	0.81	1.02	1.13	1.29	1.31	1.32	1.22	1.04	0.94	0.79	0.74
47	0.77	0.80	1.02	1.14	1.30	1.32	1.33	1.22	1.04	0.93	0.78	0.73
48	0.76	0.80	1.02	1.14	1.31	1.33	1.34	1.23	1.05	1.93	0.77	0.72
49	0.75	0.79	1.02	1.14	1.32	1.34	1.35	1.24	1.05	0.93	0.76	0.71
50	0.74	0.78	1.02	1.15	1.33	1.36	1.37	1.25	1.06	0.92	0.76	0.70
South Latitude												
5	1.06	0.95	1.04	1.00	1.02	0.99	1.02	1.03	1.00	1.05	1.03	1.06
10	1.08	0.97	1.05	0.99	1.01	0.96	1.00	1.01	1.00	1.06	1.05	1.10
15	1.12	0.98	1.05	0.98	0.98	0.94	0.97	1.00	1.00	1.07	1.07	1.12
20	1.14	1.00	1.05	0.97	0.96	0.91	0.95	0.99	1.00	1.08	1.09	1.15
25	1.17	1.01	1.05	0.96	0.94	0.88	0.93	0.98	1.00	1.10	1.11	1.18
30	1.20	1.03	1.06	0.95	0.92	0.85	0.90	0.96	1.00	1.12	1.14	1.21
35	1.23	1.01	1.06	0.94	0.89	0.82	0.87	0.94	1.00	1.13	1.17	1.25
40	1.27	1.06	1.07	0.93	0.86	0.78	0.84	0.92	1.00	1.15	1.20	1.29
42	1.28	1.07	1.07	0.92	0.85	0.76	0.82	0.92	1.00	1.16	1.22	1.31
44	1.30	1.06	1.07	0.92	0.83	0.74	0.81	0.91	0.99	1.17	1.23	1.33
46	1.32	1.10	1.07	0.91	0.82	0.72	0.79	0.90	0.99	1.17	1.25	1.35
48	1.34	1.11	1.08	0.90	0.80	0.70	0.76	0.89	0.99	1.18	1.27	1.37
50	1.37	1.12	1.08	0.89	0.77	0.67	0.74	0.88	0.99	1.19	1.29	1.41

5.3.3.2. Modified Blaney-Criddle Method

In the modified Blaney-Criddle method the relationship recommended representing mean value over the given month is expressed (Doorenbos and Pruitt, 1977) as:

$$ET_0 = C\left[P\ (0.46\ T + 8)\right] \qquad ...(5.7)$$

where, ET_0 = reference crop evapotranspiration, mm/day for the month considered; T = mean daily temperature °C over the month considered; P = mean daily percentage of total annual daytime hours (Table 5.2) for a given month and latitude and C = adjustment factor which depends on minimum relative humidity, sunshine hours and day time wind estimates.

Table 5.2. Mean daily percentage (p) of annual daytime hours for different latitudes

Latitude	North South	Jan.- July	Feb.- Aug	March- Sept.	April- Oct.	May- Nov.	June- Dec.	July- Jan.	Aug.- Feb.	Sept.- March	Oct.- April	Nov.- May	Dec.- June
60°		0.15	0.20	0.26	0.32	0.38	0.41	0.40	0.34	0.28	0.22	0.17	0.13
58°		0.16	0.21	0.26	0.32	0.37	0.40	0.39	0.34	0.28	0.23	0.18	0.15
56°		0.17	0.21	0.26	0.32	0.36	0.39	0.38	0.33	0.28	0.23	0.18	0.16
54°		0.18	0.22	0.26	0.31	0.36	0.38	0.37	0.33	0.28	0.23	0.19	0.17
52°		0.19	0.22	0.27	0.31	0.35	0.37	0.36	0.33	0.28	0.24	0.20	0.17
50°		0.19	0.23	0.27	0.31	0.34	0.36	0.35	0.32	0.28	0.24	0.20	0.18
48°		0.20	0.23	0.27	0.31	0.34	0.36	0.35	0.32	0.28	0.24	0.21	0.19
46°		0.20	0.23	0.27	0.30	0.34	0.35	0.34	0.32	0.28	0.24	0.21	0.20
44°		0.21	0.24	0.27	0.30	0.33	0.35	0.34	0.31	0.28	0.25	0.22	0.20
42°		0.21	0.24	0.27	0.30	0.33	0.34	0.33	0.31	0.28	0.25	0.22	0.21
40°		0.22	0.24	0.27	0.30	0.32	0.34	0.33	0.31	0.28	0.25	0.22	0.21
35°		0.23	0.25	0.27	0.29	0.31	0.32	0.32	0.30	0.28	0.25	0.23	0.22
30°		0.24	0.25	0.27	0.29	0.31	0.32	0.31	0.30	0.28	0.26	0.24	0.23*
25°		0.24	0.26	0.27	0.29	0.30	0.31	0.31	0.29	0.28	0.26	0.25	0.24
20°		0.25	0.26	0.27	0.28	0.29	0.30	0.30	0.29	0.28	0.26	0.25	0.25
15°		0.26	0.26	0.27	0.28	0.29	0.29	0.29	0.28	0.28	0.27	0.26	0.25
10°		0.26	0.27	0.27	0.28	0.28	0.29	0.29	0.28	0.28	0.27	0.26	0.26
5°		0.27	0.27	0.27	0.28	0.28	0.28	0.28	0.28	0.28	0.27	0.27	0.27
0°		0.27	0.27	0.27	0.27	0.27	0.27	0.27	0.27	0.27	0.27	0.27	0.27

Table 5.3. Mean daily maximum duration of bright sunshine hours (N) for different months and latitudes

North South lats.	Jan.- July	Feb.- Aug	March- Sept.	April- Oct.	May- Nov.	June- Dec.	July- Jan.	Aug.- Feb	Sept.- March.	Oct.- April	Nov.- May	Dec.- Jun
50	8.5	10.1	11.8	13.8	15.4	16.3	15.9	14.5	12.7	10.8	9.1	8.1
48	8.8	10.2	11.8	13.6	15.2	16.0	15.6	14.3	12.6	10.9	9.3	8.3
46	9.1	10.4	11.9	13.5	14.9	15.7	15.4	14.2	12.6	10.9	9.5	8.7
44	9.3	10.5	11.9	13.4	14.7	15.4	15.2	14.0	12.6	11.0	9.7	8.9
42	9.4	10.6	11.9	13.4	14.6	15.2	14.9	13.9	12.6	11.1	9.8	9.1
40	9.6	10.7	11.9	13.3	14.4	15.0	14.7	13.7	12.5	11.2	10.0	9.3
35	10.1	11.0	11.9	13.1	14.0	14.5	14.3	13.5	12.4	11.3	10.3	9.8
30	10.4	11.1	12.0	12.9	13.6	14.0	13.9	13.2	12.4	11.5	10.6	10.2
25	10.7	11.3	12.0	12.7	13.3	13.7	13.5	13.0	12.3	11.6	10.9	10.6
20	11.0	11.5	12.0	12.6	13.1	13.3	13.2	12.8	12.3	11.7	11.2	10.9
15	11.3	11.6	12.0	12.5	12.8	13.0	12.9	12.6	12.2	11.8	11.4	11.2
10	11.6	11.8	12.0	12.3	12.6	12.7	12.6	12.4	12.1	11.8	11.6	11.5
5	11.8	11.9	12.0	12.2	12.3	12.4	12.3	12.3	12.1	12.0	11.9	11.8
0	12.1	12.1	12.1	12.1	12.1	12.1	12.1	12.1	12.1	12.1	12.1	12.1

The adjustment factor is determined by using Fig. 2. Fig. 2 can be used to estimate ET_O graphically using the calculated values of [P (0.46 T+8)]. Values of [P (0.46 T+8)] is given on X-axis and the value of ET_O can be read directly from the Y axis. Relationships are presented in Fig. 5.2 for (i) minimum relative humidity (RH min) (ii) three levels of the ratio of actual to maximum possible sun shine hours (n/N) and (iii) three levels of day time wind velocity at 2 m height (U day). Values of maximum possible sunshine hours for different latitudes and months are given in Table 5.3 (Doorenbos and Pruitt, 1977). The limitations of use of the method are also mentioned in the above mentioned reference.

Calculation of mean daily ET_O should be made for periods no shorter than one month. ET_O should preferably be calculated for each calendar month for each year of record rather than by using mean temperature based on several years' record. Once ET_O has been determined, the crop coefficients (kc) can be used to determine ETcrop.

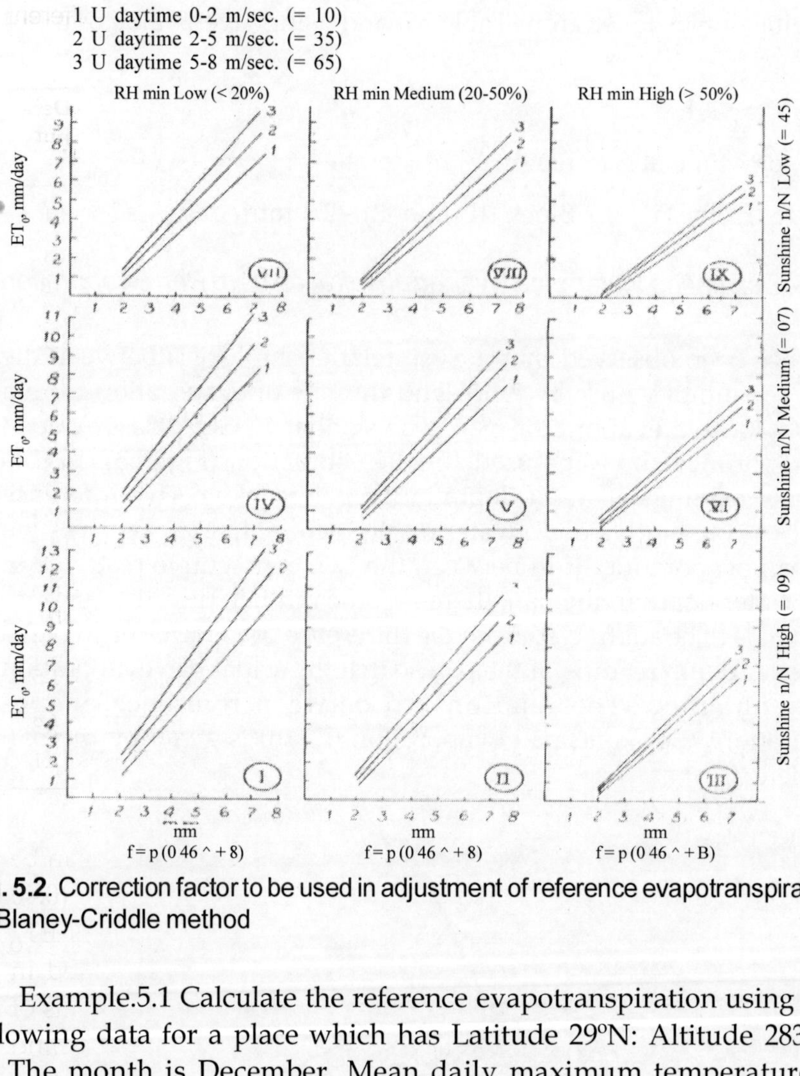

Fig. 5.2. Correction factor to be used in adjustment of reference evapotranspiration by Blaney-Criddle method

Example.5.1 Calculate the reference evapotranspiration using the following data for a place which has Latitude 29ºN: Altitude 283.89 m. The month is December. Mean daily maximum temperature = 22.96ºC and mean daily minimum temperature = 5.86ºC. RH min = 65%, day time wind velocity at 2 m height = 4 m/sec and actual sunshine hour = 9.4 hours.

Solution: T daily mean = (22. 96 + 5.86) /2 = 14.41ºC.

P from Table 5.2 for 29ºN = 0.23

P (0.46 T+8) = 0.23 (0.46 X 14.41 + 8) = 3.36 mm/day

RH min = high

Value of N is taken from Table 5.3 and using given value of n (9.4 hr) we get,

n/N = high

u_2 day time at 2 m height = moderate

ET_O from Fig. 5.2 Block III (line 2) =2.3 mm/day

5.3.3.3. Estimating Reference Evapotranspiration from Pan Evaporation Data

It has been observed that a close relationship exists between the rate of consumptive use by crops and the rate of evaporation from a properly located evaporation pan. The standard USDA class A open pan evaporimeter may be used for the estimation (Fig. 5.3). In this case daily readings of water depth in the pan is taken. The difference of the two consecutive day's water depth reading in the pan gives the daily pan evaporation. If in between the two consecutive readings of taking water depth in the pan, there is some rainfall, then this amount of rainfall is to be subtracted from the difference of the two consecutive days water depth reading of the pan so that the actual pan evaporation data is obtained. The relationship between reference evapotranspiration (ETo) and pan evaporation (Epan) is given by the pan coefficient (Kp) as:

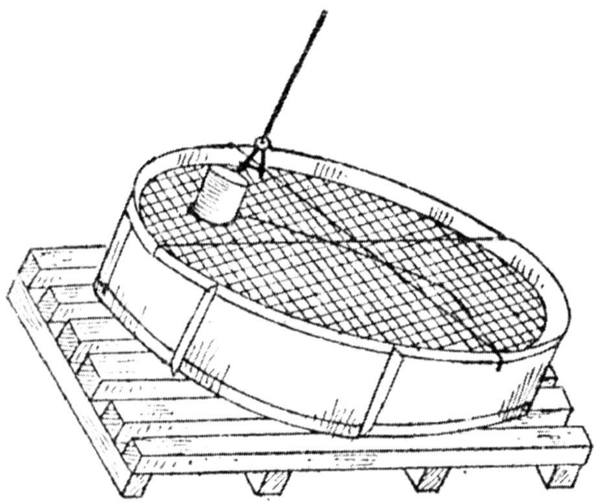

Fig. 5.3 Figure of U.S.D.A. open pan

$$ET_0 = Epan \times Kp \qquad \text{... (5.8)}$$

The value of Kp depends on environment and geographical location and also on the nature and characteristics of pan. For detail discussions on finding out the value of Kp, reference of Michael (1978) may be made.

Other methods used for computation of ETo are Penmen method, Modified Penmen method, Radiation method, Christiansen method, Energy Balance method, Oliver method, Van Bavel method, FAO Penmen-Monteith method etc. Out of these methods, FAO Penmen-Monteith method gives reasonable correct value of ETo since it considers a number of climatic data. The details of FAO Penmen-Monteith method is discussed in FAO Paper No. 56 (Allen et al., 1998).

5.4 Crop Coefficient for Estimating Actual Evapotranspiration

The values of ET_0 (reference evapotranspiration) estimated by using different methods based on climatological parameters, as described earlier, need to be adjusted for actual evapotranspiration (ETcrop) under natural field conditions. ET_0 rarely occurs in most of the irrigated field crops, with an exception of lowland rice and probably for two to three days immediately after irrigation or rains. For converting ET_0 values into *ETcrop*, suitable crop coefficients (k_c) should be evolved for different crops, soils and climatic conditions and also for different stages of growth for the same crop. Conversion of *ETo* to *ETcrop* is given as:

$$ETcrop = k_c \; x \; ET_0 \qquad (5.9)$$

The range of k_c values during different growth stages of crop is illustrated in Fig. 5.4. The value of k_c largely depends on the level of ET_O and frequency with which soil is wetted by rain and/or irrigation. The smooth curve in the figure (Fig. 5.4) presents average k_c values rather than actual sharp increase in k_c just following rain and irrigation,with a less sharp but marked decline afterwards, until the next rain or irrigation. The ratio may exceed 1.0 in areas prone to advection since the same is not reflected in ETo estimated by most of the formulae. The correct estimation of ET from ET_0, therefore, depends upon the accuracy of the K_c values adopted. Values of actual crop evapotranspiration are low during the .early stages of crop growth and increases as the plant approaches full growth period and remain fairly constant for some time. There is a general decline during the late/matured periods (Michael, 1978). The values of k_c for a particular crop at a location should be determined experimentally. If the values of k_c for particular crop and place are not available then the k_c values as reported by Doorenbos and Pruitt (1977) may be used.

The value of k_c depends on type of crop, its management practice, soil moisture condition and growth stages of the crop. The values are low during the initial and matured stage of the crop and are more during the crop development and mid season stage. During the mid season and crop development stage, the crop has more canopies, covers the soil to the maximum extent and has more developed and maximum root development. This necessitates the plants to require higher amount of water which is reflected in higher values of k_c. It is also to be remembered that the soil moisture content is a deciding factor for evaluation of k_c values.

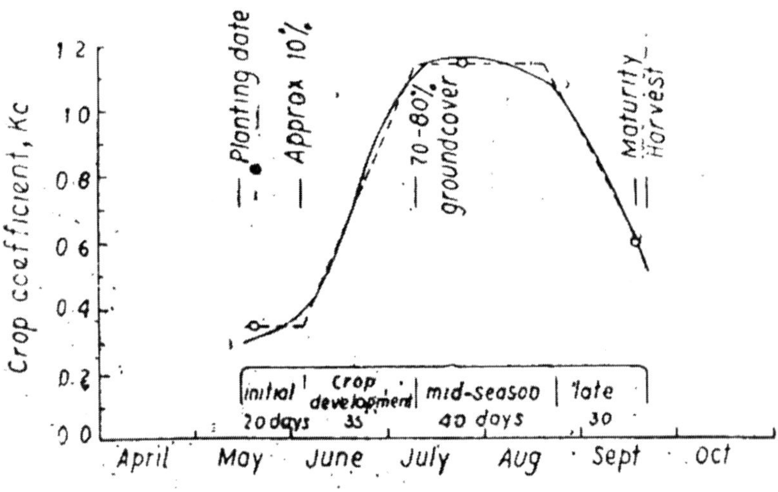

Fig. 5.4. Examples of crop coefficient, kc

Under adequate soil moisture condition the values of k_c are more and as the soil moisture decreases and the plants remain under stress condition, values of k_c become less as a result of which, the actual crop evapotranspiration decreases from the potential value. Under such stress condition, the actual crop evapotranspiration (ETcrop) is given as:

$$ET_{crop} = k_s \times k_c \times ET_0 \qquad ...(5.10)$$

where, ks = crop stress factor and other terms are defined earlier.

The value of crop stress factor is equal 1.0 under adequate soil moisture condition (optimum level) and decreases from 1.0 as the soil moisture decreases from the optimum value. Some researchers report that there is a linear decrease of actual crop evapotranspiration from the potential values as the soil moisture decreases from the optimum value whereas others opine a logarithmic decline.

Paddy crop, in general has kc values ranging from 0.9 to 1.2 under ponding and at about saturation condition. However, as the ponding water from the crop field disappears such that the unsaturated condition occurs in the fields, the value of kc linearly decreases. Jensen et al. (1993) reported that kc values of paddy decreases by 10% every day from the day of disappearance of ponded water from the field. Panigrahi (2001) reported that in paddy field under unsaturated condition, the crop stress factor is equal to the relative available soil moisture which is the ratio of current available soil moisture in the crop effective root zone to the saturated moisture content in the crop effective root zone. The values of crop stress factor results in decline of actual crop evapotranspiration and hence yield. It is to be remembered that the stress condition in the crop during the critical growth stages has pronounced effect on decreasing the yield than that occurring in other stages. Detail discussions on effect of stress factor on crop at different growth stages along with the values at various stages of different crops are available in FAO Paper No. 33 (Doorenbos and Kassam, 1979).

Question Banks

Q1. What are the multiple uses of water of a farm pond?

Q2. What are the factors considered for computing the total crop water demand of a farm pond service area?

Q3. What is the normal water demand of large livestock like horse and cattle and small livestock like pig and sheep?

Q4. How can the total water demand of the farm pond be determined?

Q5. Define the following terms.

(i) Consumptive use (ii) Water requirement (iii) Evapotranspiration (iv) Duty and (v) Delta.

Q6. Briefly explain the energy balance method to estimate the crop water requirement.

Q7. With label diagram, describe how the evapotranspiration of crop is measured by the lysimeters.

Q8. Which type of lysimeter gives more correct measured values of crop evapotranspiration? Describe how this lysimeter is used to measure the crop evapotranspiration.

Q9. Describe how the soil moisture depletion study is helpful to compute the crop evapotranspiration.

Q10. Profile soil moisture content of a potato field in 60 cm effective root zone depth are measured as follows:

Profile layer, cm	Soil moisture content,%		Apparent Specific gravity, gm/cc
	Beginning of irrigation cycle	End of irrigation cycle	
0-15	20	12	1.60
15-30	22	13	1.58
30-45	24	15	1.59
45-60	25	15	1.60

If effective rainfall within the irrigation cycle is 20 mm, and percolation from the effective root zone of the crop is 11 mm, estimate the water use from the crop root zone within the irrigation cycle.

Q11. A field is sown with ground nut. The profile soil moisture content of the field at sowing and at harvest time are measured and presented in table as below.

Profile layer, cm gm/cc	Soil moisture content,%		Apparent Specific gravity,
	At sowing	At harvest	
0-15	22	10	1.62
15-30	24	11	1.60
30-45	25	13	1.59
45-60	25	15	1.60

If the crop is supplied with 20 cm irrigation, effective rainfall in the crop growing season is 12 cm and there is no percolation loss from the crop effective root zone in the entire season, then estimate the seasonal water requirement of the crop.

Q12. Write the general water balance model of crop field and explain the different terms. How can the model be used to predict the soil moisture content in different soil layers of the crop filed?

Q13. Write down the different steps involved in estimating the reference crop evapotranspiration by Thronwaite method.

Q14. Following data are recorded for a particular place having latitude 290N. Estimate the daily reference crop evapotranspiration for the month of January using the data.

Month	Jan	Feb	Mar	Apr	May	Jun	Jul	Aug	Sep	Oct	Nov	Dec
Mean temperature,°C	13.0	15.9	16.5	23	29.8	30.4	29.0	28.6	25.7	20.6	18.6	12.1

Q15. Write down the different steps used to compute the reference evapotranspiration by the modified Blaney-Criddlee method.

Q16. Calculate the reference evapotranspiration with using the following data.

Place is Chiplima having 220N latitude and altitude is 205 m period is month December. The mean maximum temperature is 23oC and mean minimum temperature is 6oC. The place has actual sunshine hour (average monthly) is 9 hr. The day wind velocity at 2 m height is 3.5 m/s and the mean minimum relative humidity is 54%.

If the crop coefficient value is 1.05 then find out the actual crop evapotranspiration.

Q17. The water level in a pan evaporimeter on a day at 7 AM is 20 cm below the rim of the pan and that on the next day is 19.2 cm. If there is no rainfall within the time both readings are taken then find out the pan evaporation. Had there been 5 mm rainfall then what would have been the value of pan evaporation? Using pan coefficient value as 0.8, find out the reference evapotranspiration in both the cases.

Q18. What do you mean by crop coefficient and how it is useful in estimation of actual crop evapotyranspiration?

Q19. What are the factors that influence values of the crop coefficient?

Q20. What do you mean by crop stress factor? How does it affect the actual crop evapotranspiration?

Q21. Find out the seasonal actual crop evapotranspiration of a crop using the data given in tabular form as below.

Crop growth stage	**Values of reference evapotranspiration, cm**	**Values of crop coefficient**	**Values of crop stress factor**
Initial	20	0.8	0.5
Crop development	25	1.0	0.9
Mid season	30	1.2	0.9
Maturity	18	0.9	1.0

References

Allen, R.G., Pereira , L.S., Raes, D. and Smith, M. 1998. Crop Evapotranspiration: Guidelines for Computing Water Requirements. Irrigation and Drainage Paper 56, FAO, Rome, pp 300.

Doorenbos, J. and Kassam, A.H. 1979. Yield Response to Water. Irrigation and Drainage Paper 33, FAO, Rome.

Doorenbos, J. and Pruitt, W.O. 1977. Guidelines for Predicting Crop water Requirements. Irrigation and Drainage Paper 24, FAO, Rome

Jensen, J.R., Mannan, S.M.A. and Uddin, S.M.N. 1993. Irrigation equirement of transplanted monsoon rice in Bangladesh. Agricultural Water Management, Vol. 23(3):199-212.

Lenka, D. 2001. Irrigation and Drainage. Kalyani Publishers, New Delhi.

Michael, A.M. 1978. Irrigation Theory and Practices. Vikash Publishing House Pvt. Ltd., New Delhi.

Misra, R.D. and Ahmed, M. 1987. Manual on Irrigation Agronomy, Oxford & IBH Publishing Co. Pvt. Ltd., New Delhi, pp. 412.

Odhiambo, L.O. and Murty, V.V.N. 1996. Modeling water balance components in relation to field layouts in low land paddy fields. I. model development, Agricultural Water Management, Vol. 30(2): 185-199.

Panigrahi, B. 2001. Water Balance Simulation for Optimum Design of On-Farm Reservoir in Rainfed Farming System. Ph.D. Thesis, Indian Institute of Technology, Kharagpur.

Thornthwaite, C.W. 1948. An approach towards a rational classification of climate. Georgr. Rev.: 55-94.

Chapter 6

Dimensions of Farm Pond

6.1 Introduction

Farm ponds used in farmers' fields are generally square shaped or rectangular shaped pyramidal structures. It has a regular side slope, depth, and top and bottom dimensions. It may be excavated/dugout pond or embankment pond. In excavated pond, the pond is constructed below the ground level. Water is collected in the pond by gravity from the adjacent catchment. In embankment type, the pond is partly above the ground and partly below. This type of pond has higher storage capacity than the dugout type pond. This type of pond has a berm and dyke/embankment around the pond. The embankment has some side slope. The pond has bottom dimensions, dimensions at the ground level as well as some dimensions at the top above the embankment. In this chapter, various dimensions of pond like depth, side slope of the pond, as well as the various dimensions at bottom, ground and top level will be discussed.

6.2 Depth

Depth of the farm pond depend on the climatic condition. Depth of the pond varies from 2.0 m in wet climatic condition to as high as 5.5 m in arid climatic condition. For the same volume of water stored, deeper the pond, lesser is the area occupied by the pond and also lesser are the evaporation losses. However, it should be kept in mind that a large depth of the pond requires large pond area from stability point of view. When the construction of the pond is done manually, any increase in depth beyond 3 m becomes uneconomical sine the cost increase out of the proportion to the volume of excavation. It also becomes uneconomical and difficult for lifting devices operated with human and animal power. A depth of 2-3 m is optimum for ponds in black clay soils. In general, the depth of the pond in coarse textured soil is less than that in the fine textured soil from stability point of view (Senapati, 2008).

6.3 Side Slope

Suitable side slope of the pond is assumed. The side slope is decided based on the angle of repose of the sub-soil. Generally, firm clay soils have an angle of repose of 45^0. It also depends on the type of the soil. For coarse textured soil, the slope is flatter and for the heavy soil it is steeper. The slope is generally expressed as ratio of horizontal length to vertical length. The slope may vary from 1:1 for stable heavy textured soil like clay to 3:1 for less stable coarse textured soil like sandy soil.

6.4 Computation of Dimensions

After selecting the depth and side slope of the pond, the dimensions are calculated for a given volume/storage of pond. Fig. 6.1 shows a typical dug out square shaped pond. The dimensions of this type of pond can be determined once the depth, side slope as well as the storage volume of the pond is known. The bottom dimension of the pond is given as (Belgaumi et al., 1997):

$$l = \frac{\sqrt{3\ V - d^2 \times St^2}}{3\ d} - d \times St \qquad ...(6.1)$$

where l = pond length at bottom, m; V = volume of farm pond, m^3; d = depth of the pond below ground level or above the bottom level, m and St = side slope of the pond (horizontal: vertical). Since, the pond is square shaped; the bottom area can be obtained by squaring the value obtained as above.

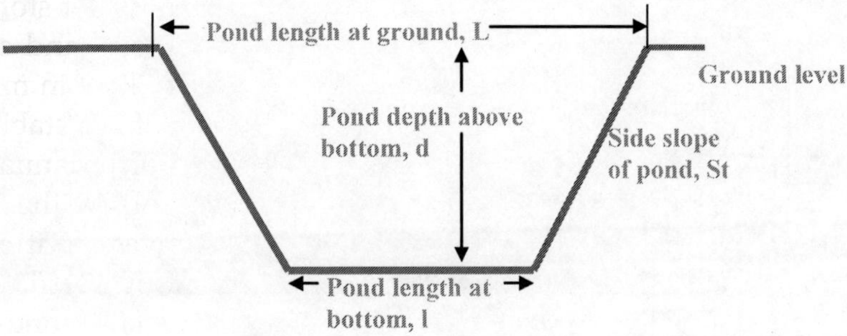

Fig. 6.1. Cross sectional view of square size dug out farm pond

Top area: The length of side of the pond at the top i.e. pond length at the ground level is calculated as:

$$L = l + 2d \times St \qquad ...(6.2)$$

where, L = length of pond at the top or ground level and other terms of the above equation are defined earlier. Once the values of l, d and St are known, then one can compute the length of the pond at the top. Top area of the pond can be calculated by squaring the length of the pond at the top.

Fig. 6.2 shows a typical square size pyramidal embankment type farm pond (which is partly dug out and partly above the ground level) with various terminologies.

Let the length and width of the pond at the top be Ls and Bs, respectively (Fig. 6.2). For a square shaped pyramidal pond, Ls = Bs. Now use the following equations to find out the dimensions of the pond.

$$L = L_s - 2\left[2\left\{H_e.S_e + T_w + B_w\right\}\right] \qquad ...(6.3)$$

$$B = B_s - 2\left[2\left\{H_eS_e + T_w + B_w\right\}\right] \qquad ...(6.4)$$

$$l = L - 2\left[d.S_l\right] \qquad ...(6.5)$$

$$b = B - 2(d.S_l) \qquad ...(6.6)$$

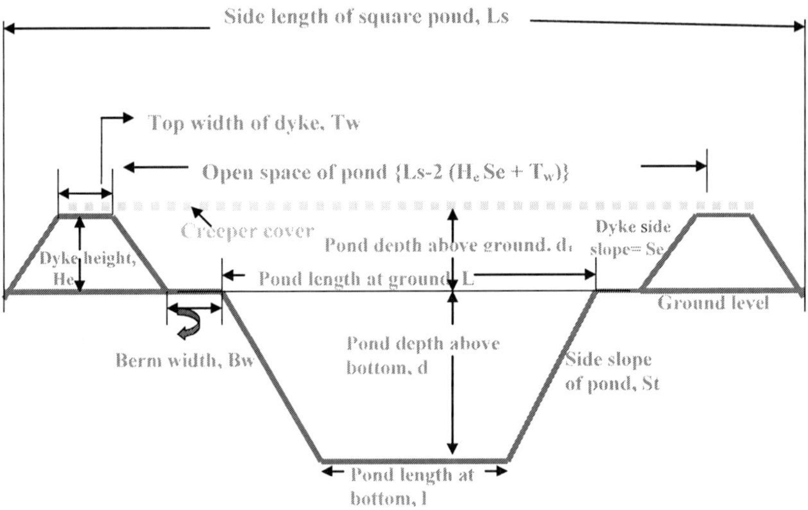

Fig. 6.2. Cross sectional view of embankment type farm pond

The volume of the pond storage when the water depth is below the ground level is computed by the prismoidal formula as (Panigrahi, 2001):

$$V_1 = \left[l.b + 4\{l + dS_l\}\{b + d.S_l\} + \{l + 2\,d.S_l\}\{b \;\; +2d.S_t\}\right]\frac{d}{6} \qquad ...(6.7)$$

The volume of the pond storage when the water depth is above the ground level is computed by the prismoidal formula as (Panigrahi, 2001):

$$V_2 = V_1 + [(L + 2B_w)(B + 2B_w) + 4\{L + 2B_w + d_1.S_e\}\{B + 2B_w + d_1.S_e\} +$$

$$\{L + 2B_w + 2.d_1.S_e\}\{B + 2B_w + 2.d_1.S_e\}]\frac{d_1}{6} \qquad ...(6.8)$$

In Eqns. (6.3) to (6.8) the various terms are: L = length of the pond at ground level, m; B = width of the pond at ground level, m; Bw = berm width of the pond, m; Tw = top width of the dyke of the pond, m; He = height of the dyke, m; Se = slope of the dyke (horizontal : vertical); l = bottom length of the pond, m; b = bottom width of the pond, m; d = Depth of water in pond above bottom level, m; St = side slope of the pond (horizontal : vertical); V_1 = Volume of pond water storage at depth of water d, m^3; d_1 = Depth of water in pond above the ground level, m; V_2 = Volume of pond water storage at depth of water d_1, m^3 and other terms are as defined in Fig. 6.2.

At the first instance, the values of Se, St, Tw, Bw, He, d_1 and d are assumed. Volume of pond for which dimensions are to be calculated is known. A certain value of Ls and Bs are assumed and using the above mentioned equations (Eqns. 6.3 to 6.8), the volume of the pond is calculated. If the volume so calculated is equal to the given volume, then dimensions of the pond assumed remains correct, otherwise the values of Ls and Bs are changed and again the volume is calculated. Thus, several interactions are needed to find out the dimensions of the pond.

6.5 Calculating Water Volume of Pond

It is often necessary to know the volume of a pond. If the pond is used as a source for irrigation, then the pond volume limits the amount of water available over a short period of time. Pond volume is not only important to assess the amount and frequency of irrigation to crops, it is also used to estimate the amount of recharge from the under ground reservoir. Moreover, pond volume is useful for non-farm uses. It is also used to asses the amount of water supplied to a given industry. Calculating the volume of a pond requires the shape of the pond as well as requires measuring both the pond surface area, and depth. In this section, the volume of pond for various shapes is discussed.

Estimation of depth of the pond is the vital component to determine the volume of the pond. Water depth of a pond can be measured either by placing a graduated scale at the centre of the pond. The depth is usually measured by taking the readings of the scale. However, for large ponds having higher depth, water depth can be easily measured from a boat along two perpendicular transects. One simple approach is to use a weighted line to measure depth at a number of evenly-spaced locations along each transect. To compute the average depth, the depth measurements should be added and then the addition is divided by the number of measurements.

There are some sophisticated approaches to measure the pond depth. A hand -held sonar device is used now commercially to measure the pond depth. The hand -held sonar device can measure depth up to 200 ft. The commercial model Polarvision by Strike Master which is often used is about the size of large flash light and can be used to measure water depth from a boat (Coche *et al.,* 1995).

There are several approaches used to measure the volume of a pond. The simplest approach to measure the volume of a rectangular pond (Fig. 6.3) is given as:

$$V = l \times b \times d \quad \text{...(6.9)}$$

where, V = volume of the pond, l = length of the pond, b = width of the pond and d = average depth of the pond.

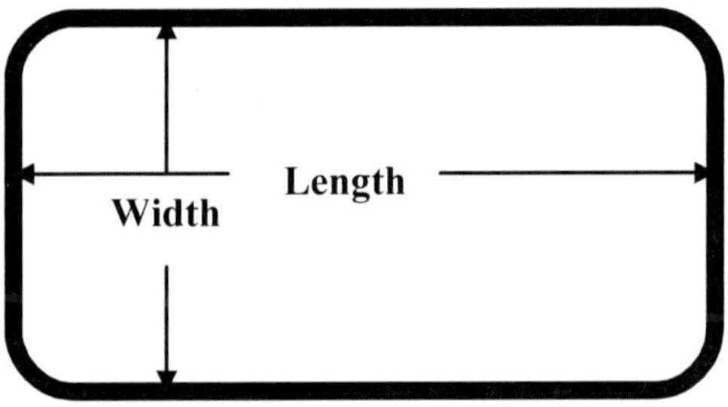

Fig. 6.3. Rectangular pond

In calculation of volume of the pond by the above mentioned formula the limitation is that the pond has steep side slope i.e. it is a vertical ditch.

For a circular pond (Fig. 6.4) with steep side slopes, volume of the pond is given as:

$$V = \pi \times R^2 \times d \quad \text{...(6.10)}$$

where, R = radius of the pond and d = average depth of the pond.

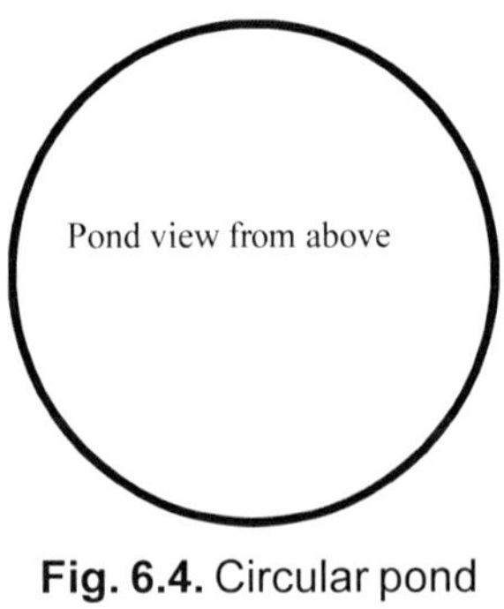

Fig. 6.4. Circular pond

For a pond with conical depth profile and having circular shape (Fig. 6.5), the volume is expressed (Coche et al., 1995) as:

$$V = 1.047 \times R^2 \times D \quad ...(6.11)$$

where, V = volume of the pond, R = radius of the pond D = maximum depth of the pond.

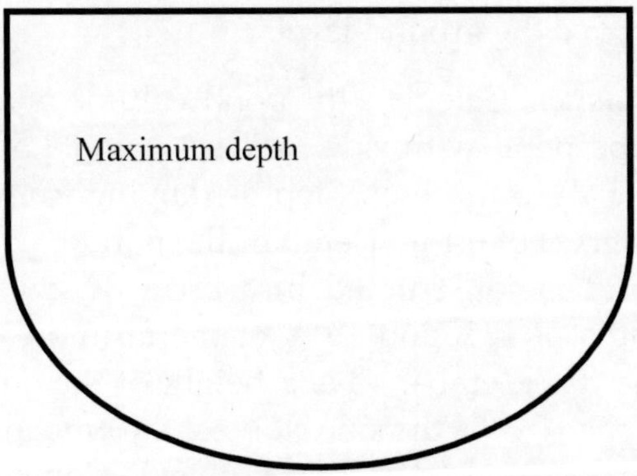

Fig. 6.5. Conical shape pond

6.5.1. Water Volume of a Square Shaped Dugout Pond

After determining the side slope, and dimensions of the pond, the volume of a dug out pond can be computed by using several formulae. Some of the commonly used formulae to determine the volume are mentioned as below (Murty, 1998).

Prismoidal Formula: As per this formula the volume is given as:

$$V = (A + 4B + C) \times d/6 \quad ...(6.12)$$

where, V = volume of pond, m^3; A = area of pond at the top i.e. at ground level, m^2; B = area of the pond at mid depth, m^2; C = area of the pond at bottom level, m^2 and d = average depth of the pond, m.

End Area Method: By this method, the volume is computed as:

$$V = (A_1 + A_2) \times d/2 \quad ...(6.13)$$

where, A_1 and A_2 are the area of the pond at bottom and top level, respectively, m^2 and other terms are as defined earlier. The volume of the pond is given in m^3 and average depth is in m.

6.5.2. Water Volume of a Square Shaped Embankment Pond

Water volume of a square shaped embankment type pond (Fig. 6.2) computed by prismoidal formula is mentioned in Eqns. (6.7) and (6.8) with various terms of these equations being defined earlier. We can compute the pond water volume when the depth of water in the pond is below the ground level by using Eq. (6.7) and the pond volume can be computed by using Eq. (6.8) when the depth of water in the pond is above the ground level.

Sample Calculation: Calculate the pond volume of a square shaped embankment type pond with side slopes 1:1 and 1.5:1 (side slope is horizontal : vertical) for different depths ranging from 0.1 to 2.3 m at an increasing interval of 0.1 m. Depth of the pond below ground level is 2 m. The pond is constructed in a farm of size 800 m^2 and it constitutes an area of 12% and 10 % of the farm areas, respectively. The width of the dyke/embankment, height of the embankment and berm width are each 0.3 m, the side slope of the embankment is 1:1 for both 12 and 10% size of pond and for both side slopes 1:1 and 1.5:1 of the pond.

Solution:

Dimensions for 12% farm size: The top surface area of the pond is 12% of farm area and the farm area is given as 800 m^2. Hence the top surface area of the pond is 96 m^2.

Since the pond is square shaped, length = width of each side of the pond at the top (above the embankment level) (Ls) = 9.8 m.

Since the embankment has side slope 1:1, top width = 0.3 m, height = 0.3 m and berm width = 0.3 m, the length = width of the pond at the ground level (L) = 7.4 m (Eq. 6.3)

For 2 m depth of the pond below ground level and 1:1 and 1.5:1 side slope of the pond, the length = width of the pond at the bottom (l) is computed as 3.4 and 1.4 m, respectively (Eq. 6.5).

Dimensions for 10% farm size: The top surface area of the pond is 10% of farm area and the farm area is given as 800 m^2. Hence the top surface area of the pond is 80 m^2.

Since the pond is square shaped, length = width of each side of the pond at the top (above the embankment level) (Ls) = 8.95 m.

Since the embankment has side slope 1:1, top width = 0.3 m, height = 0.3 m and berm width = 0.3 m, the length = width of the pond at the ground level (L) = 6.55 m (Eq. 6.3)

For 2 m depth of the pond below ground level and 1:1 and 1.5:1 side slope, the length = width of the pond at the bottom (l) is computed as 2.55 and 0.55 m, respectively (Eq. 6.5).

The volume of the pond for both 10 and 12% farm sizes and side slopes 1:1 and 1.5:1 is computed by using Eq. (6.7) for water depth up to 2 m and then for depths beyond 2 m, the volume is computed by Eq. (6.8) and is presented in Table 6.1.

Table 6.1. Computation of volume of embankment type pond (m^3) for different side slope and size for a farm area of 800 m^2

Pond depth, m	Volume at 12% size of farm pond		Volume at 10% size of farm pond	
	Side slope 1:1	Side slope 1.5:1	Side slope 1:1	Side slope 1.5:1
0.1	1.23	0.23	0.70	0.04
0.2	2.59	0.51	1.52	0.12
0.3	4.12	0.88	2.45	0.23
0.4	5.80	1.32	3.50	0.38
0.5	7.65	1.85	4.69	0.59
0.6	9.67	2.47	6.03	0.87
0.7	11.88	3.20	7.51	1.21
0.8	14.28	4.04	9.15	1.63
0.9	16.88	5.00	10.96	2.14
1.0	19.69	6.09	12.94	2.74
1.1	22.72	7.32	15.10	3.44
1.2	25.97	8.69	17.45	4.25
1.3	29.45	10.21	20.00	5.18
1.4	33.17	11.89	22.76	6.24
1.5	37.14	13.74	25.73	7.43
1.6	41.37	15.77	28.92	8.76
1.7	45.85	17.97	32.34	10.24
1.8	50.62	20.38	36.00	11.88
1.9	55.66	22.98	39.91	13.69
2.0	60.99	25.79	44.07	15.67
2.1	67.54	32.34	49.33	20.93
2.2	74.43	39.23	54.88	26.48
2.3	81.66	46.46	60.73	32.33

6.6 Top Water Surface Area of Farm Pond

After the dimensions of the farm pond are found out, it is easy to compute the water spread area otherwise called as top water surface area. The water spread area is helpful in estimation of water losses in the form of evaporation from the pond. The water spread area of a pond depends on the shape and size of the pond, side slope of the pond as well as depth of water in the pond at any instant of time. If the side slope of the pond is more i.e. it is flatter, then the water spread area of the pond for a given size and shape of the pond at a particular depth of water in the pond will be more than that at steeper slope. In this section the water spread area of dug out and embankment type pond is discussed.

Dugout Pond

For a dugout pond (Fig. 6.1), the water spread area otherwise called as top water surface area for any depth of water "d " above bottom level of pond is:

$$TS_1 = (l + 2.d.S_1)(b + 2.d.S_1) \quad \text{...(6.14)}$$

Embankment Type Pond

For an embankment type pond (Fig. 6.2), the water spread area for any depth of water "d" above bottom level in pond, TS_1 is given as above (Eq. 6.14)

However, when the depth of water in the pond is above the ground level, d_1, then the water spread area is given (Panigrahi, 2001) as:

$$TS_2 = (L + 2.B_w + 2.d_1 S_e)(B + 2B_w + 2d_1.S_e) \quad \text{...(6.15)}$$

The various terms of the above mentioned two equations (Eq. 6.14 and 6.15) are defined earlier in the section "computation of volume" and also in the Fig. 6.2.

Sample Calculation: Calculate the water spread area of a square shaped embankment type pond with side slopes 1:1 and 1.5:1 (side slope is horizontal : vertical) for different depths ranging from 0.1 to 2.3 m at an increasing interval of 0.1 m. Depth of the pond below ground level is 2 m. The pond is constructed in a farm of size 800 m^2 and it constitutes an area of 12% and 10 % of the farm areas, respectively. The width of the dyke/embankment, height of the embankment and berm width are each 0.3 m, the side slope of the embankment is 1:1 for both 12 and 10% size of pond and for both side slopes 1:1 and 1.5:1 of the pond.

Solution: Dimensions of the pond for both 10 and 12% size for a farm area of 800 m^2 with 1:1 and 1.5:1 side slope of the pond is same as computed for the sample calculation of volume of the pond. Then using Eq. (6.14), we estimate the water spread area of the pond for both 10 and 12% pond sizes and 1:1 and 1.5:1 pond side slope for various depths ranging from 0.1 to 2 m and for depths beyond 2 m, Eq. (6.15) is used. The computations of water spread areas are presented in Table 6.2.

6.7 Wetted Surface Area of the Pond

After finding out the dimensions of the pond, we can calculate the wetted surface areas of the pond for various depths. The wetted surface area depends on the size and shape of the pond as well as the depth and side slope of the pond. For a flatter slope, the wetted surface area of the pond will be more compared to the steeper side slope. The wetted surface area helps in the computation of seepage losses in the pond. As the depth of the pond varies depending on the inflow and out flow to and from the pond, the pond storage changes everyday. The seepage losses occur from the ponds everyday and this seepage occurs through the sides and bottoms of the pond. The magnitudes of the seepage loss depend on the wetted surface area of the pond. Computational procedure of wetted surface area as discussed in this section will help in quantifying the seepage loss from the pond and deciding the water balance of the pond.

Dugout Pond: For a dug out pond (Fig. 6.1), the wetted surface area (WS_1) for any depth of water d above bottom level of pond is (Panigrahi, 2001):

$$WS_1 = l.b + 2.(l + d.\ S_t).(d^2 + d^2.S_l^{\ 2})^{\frac{1}{2}}. + 2.(b + dS_t).(d^2 + d^2.S_l^{\ 2})^{\frac{1}{2}} \quad \text{... (6,16)}$$

Embankment Type Pond: For an embankment type pond (Fig. 6.2), the wetted surface area for any depth of water d above bottom level in pond, WS_1 is given as above (Eq. 6.16).

However, when the depth of water in the pond is above the ground level, d_1, then the wetted surface area (WS_2) is given (Panigrahi, 2001) as:

$$WS_2 = WS_1 + 2(B_w.L) + 2(d_1^{\,2} + d_1^{\,2}.S_e^{\,2})^{\frac{1}{2}}.(L + d_1\ S_e) + 2(d_1^{\,2} + d_1^{\,2}.S_e^{\,2})^{\frac{1}{2}}.(B + d_1\ S_e) + 2(B_w.B) \quad \ldots (6,17)$$

Table 6.2. Computation of water spread area of embankment type pond for different side slope and size for a farm area of 800 m^2

Pond depth, m	Water spread area, m^2			
	12% size of farm pond		10% size of farm pond	
	Side slope 1:1	Side slope 1.5:1	Side slope 1:1	Side slope1:1 1.5:1
0.1	12.96	2.89	7.56	0.72
0.2	14.44	4.00	8.70	1.32
0.3	16.00	5.39	9.92	2.10
0.4	17.64	6.76	11.22	3.06
0.5	19.36	8.41	12.60	4.20
0.6	21.16	10.24	14.06	5.52
0.7	23.04	12.25	15.60	7.02
0.8	25.00	14.44	17.22	8.70
0.9	27.04	16.81	18.92	10.56
1.0	29.16	19.36	20.70	12.60
1.1	31.36	22.09	22.56	14.82
1.2	33.64	25.00	24.50	17.22
1.3	36.00	28.09	26.52	19.80
1.4	38.44	31.36	28.62	22.56
1.5	40.96	34.81	30.80	25.50
1.6	43.56	38.44	33.06	28.62
1.7	46.24	42.25	35.40	31.92
1.8	49.00	46.24	37.80	35.40
1.9	51.34	50.41	40.32	39.06
2.0	54.76	54.76	42.90	42.90
2.1	67.24	67.24	54.02	54.02
2.2	70.56	70.56	57.00	57.00
2.3	73.96	73.96	60.06	60.06

The various terms of the above mentioned two equations (Eq. 6.16 and 6.17) are defined earlier in the section "computation of volume" and also in the Fig. 6.2.

Sample Calculation: Calculate the wetted surface area of pond for 12% and 10% of farm area of 800 m^2 with 1.5:1 and 1:1 side slope of the pond for various depths ranging from 0.1 to 2.3 m with incremental depth of 0.1 m. Other dimensions of the pond remain same as mentioned above in the sample calculation of the volume of pond.

Solution: Using the same dimensions of the ponds as computed in the sample calculation of volume of pond, the wetted surface areas were computed by using Eqns. (6.16) and (6.17) and the values are presented in Table 6.3.

Table 6.3. Computation of wetted surface area of embankment type pond for different side slopes and sizes for a farm area of 800 m^2

Pond depth, m	Wetted surface area, m^2			
	12% size of farm pond		10% size of farm pond	
	Side slope 1:1	Side slope 1.5:1	Side slope 1:1	Side slope 1.5:1
0.1	13.54	3.08	8.00	0.81
0.2	15.63	4.41	9.61	1.53
0.3	17.84	5.96	11.34	2.47
0.4	20.16	7.73	13.18	3.62
0.5	22.59	9.71	15.13	4.99
0.6	25.14	11.91	17.19	6.58
0.7	27.80	14.33	19.37	8.38
0.8	30.57	16.96	21.66	10.40
0.9	33.45	19.81	24.07	12.63
1.0	36.45	22.87	26.59	15.09
1.1	39.56	26.15	29.21	17.75
1.2	42.79	29.65	31.96	20.64
1.3	46.12	33.36	34.82	23.74
1.4	49.57	37.29	37.78	27.06
1.5	53.14	41.44	40.87	30.59
1.6	56.81	45.80	44.00	34.34
1.7	60.60	50.38	47.37	38.31
1.8	64.51	55.18	50.80	42.49
1.9	68.52	60.19	54.33	46.89
2.0	72.65	65.42	57.93	51.50
2.1	85.77	78.54	69.59	63.12
2.2	90.13	82.36	73.47	67.00
2.3	94.60	87.37	77.45	70.98

6.8 Development of Nomograph of Pond Sections

For a given size and shape of pond with constant side slope, volume of stored water in the pond, wetted surface area and water spread area otherwise called as top water surface area for different depths can be computed and then a nomograph can be developed relating the volume of stored water -depth of pond, depth-wetted surface area, wetted surface area-water spread area and water spread area-depth. Such nomographs can be developed for other sizes and shapes of the farm pond and for different size of farm areas and side slopes of the ponds. These nomographs help in computation of volume of stored water, seepage losses and evaporation losses in the pond at any depth of water in the pond. A sample nomograph for 1:1 pond side slope and square size embankment type pond with pond sizes varying from 6 to 20% of the farm area of 800 m^2 is presented in Fig. 6.6 (Panigrahi *et al.*, 2007). All the pond sizes have same side slope 1:1 and have berm width 0.30 m, side slope of the dyke/embankment 1:1 and have top width and height of the embankment 0.30 m each, respectively.

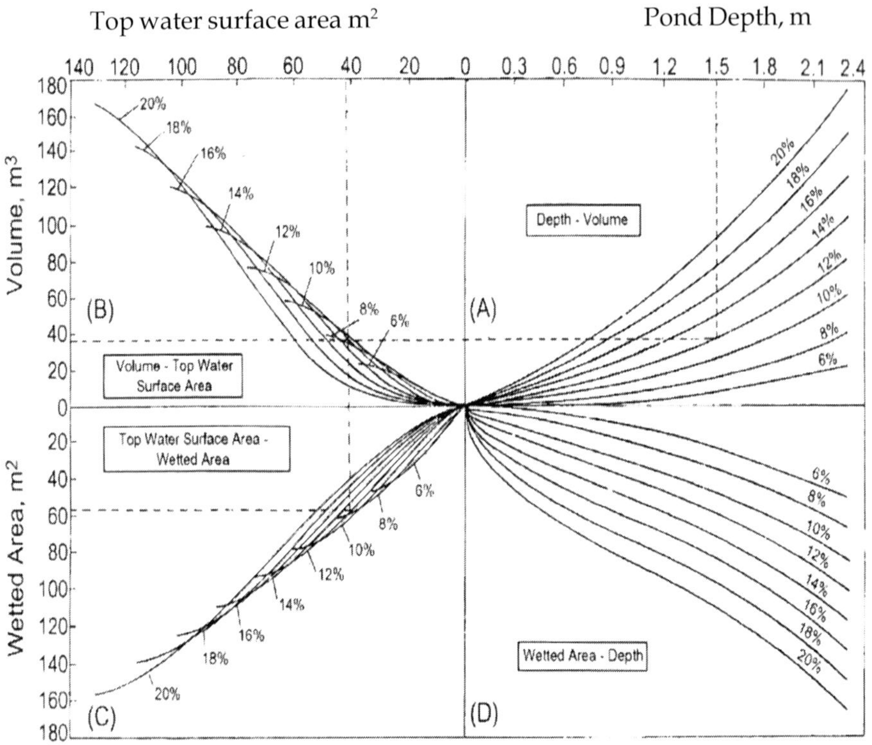

Fig.6.6. Nomograph relating various sections of the pond for different pond sizes

Question Banks

Q1. How is the depth of a farm pond decided?

Q2. What should be the side slope of a farm pond?

Q3. Draw the cross sectional view of a square size dug out pond and write down how the dimensions as well as the top water surface area of the pond can be determined.

Q4. A square size dug out farm pond has side slope 1: 1. The depth of the pond is 2 m and the bottom dimensions are 10 m x 10 m. Find out the top dimensions of the pond, volume of pond, water spread area and wetted surface area of the pond?

Q5. Draw a well labeled diagram for the cross sectional view of a square size embankment type farm pond. How will you find out the various dimensions of this pond?

Q6. Write the formula to determine the volume of pond storage in an embankment type farm pond when the water depth in the pond is below as well as above the ground level.

Q7. Write the formula to find out the volume of pond for steep side slope for the following cases.

(i) Rectangular pond (ii) Circular pond and (iii) Conical pond with circular shape.

Q8. Write the formula to compute the volume of a farm pond by prismoidal formula and end area method.

Q9. Compute the volume of a square shaped embankment type farm pond with side slope 1.5:1 (horizontal: vertical) for 2.2 m depth of pond storage. Depth of the pond below ground level is 2 m. The pond is constructed in a farm of size 1000 m^2 and it constitutes an area of 10% of the farm area. The width of the dyke/ embankment, height of the embankment and berm width are each 0.3 m, the side slope of the embankment is 1:1. Assume any other data not given in problems.

Q10. Estimate the wetted surface area and top water surface area of a farm pond constructed in a farm occupying 12% of the farm area (farm area = 1000 m^2). Depth of water in the pond is 2.1 m and the depth of pond below the ground level is 2 m. The width of the dyke/embankment, height of the embankment and berm width are each 0.3 m, the side slope of the embankment is 1:1. Assume any other data not given in problems.

Q11. Using the nomograph (Fig. 6.6), find out the volume, wetted surface area and top water surface area of a squre size embankment farm pond with 1:1 side slope of the pond. The pond has 2 m depth and occupies 10% farm area for a total farm area of 800 m^2.

References

Belgaumi, M.I., Itnal, C.J. and Radder, C.D. 1997. Farm Ponds. Technical series 4, Publication Centrte, University of Agricultural Sciences, Dharwad, pp. 47.

Coche, A.G., Muir, J.F. and Laughlin, T. 1995. Simple Methods for Aquaculture Pond Construction for Freshwater Fish Culture Building Ponds, FAO of the United Nations, Rome, pp. 355.

Murty, V.V.N. 1998. Land and Water Management Engineering. 2nd Edition, Kalyani Publishers, New Delhi, pp. 586.

Panigrahi, B. 2001. Water Balance Simulation for Optimum Design of On-Farm Reservoir in Rainfed Farming System. Ph.D. Thesis, Indian Institute of Technology, Kharagpur.

Panigrahi, B; Panda, S.N; and Mal, B.C. 2007. Rainwater Conservation and Recycling by Optimal Size On-Farm Reservoir. Resources, Conservation and Recycling. (Elsevier Publication), Vol. 50 (4): 459-474.

Senapati, S.C. 2008. Practical Manual on Soil and Water Conservation Engineering. College of Agricultural Engineering and Technology, Orissa University of Agriculture and Technology, Bhubaneswar, India.

Chapter 7

Optimum Size of Farm Pond

7.1 Introduction

The practice of rainwater storage in the pond is quite common but still there is a need to improve its design. Most of the ponds used for rainwater storage are designed on thumb rule: either they are under or over designed. An over designed farm pond can meet the supplemental irrigation needs of the crop but it may not be economical. On the other hand, an under designed pond though economical, may not be able to meet the irrigation needs of the crops at all times. It is reported that in some case up to 20% of the farm area is used for construction of farm pond. These farm ponds are found not to be optimal. Although these ponds are able to meet the crop water demand at most of the stress periods, but considering the economics, they are observed not to be optimal. The big size farm ponds need large areas to be wasted for the digging of the structures. They need huge investment for their construction and lining. Hence, proper scientific studies are needed so that its storage becomes adequate to save the crops from the adverse effects of the drought in monsoon and to assure the possibility of growing a second crop in the winter and at the same time, it is economical also. Simulation studies by daily water balance models of both the crop field and the pond help to understand the availability of water in the pond to meet the supplemental irrigation demand of the crops.

7.2 Factors Affecting Size of Farm Pond

Design of optimum size of farm pond depends on the crop and cropping system followed. But it is advisable to design a pond such that it can support supplemental irrigation to at least two crops. Since, rice is the primary crop of most of the farmers, the design may be done for rice based cropping system. The second crop in the cropping system should be a low water requiring short duration crop like mustard, pulses or oil seed crop. Sizing of the farm pond can also be done for other cropping systems like ground nut - mustard, ground nut-vegetables, black gram/green gram - sesamum and many other cropping system prevalent to the agro-climatic condition of the area where pond is to be constructed. Sizing of the pond also depends on the location of the pond. If the pond is so located that runoff from the catchment (at upstream end of the pond) is collected in the pond and utilized as supplemental irrigation to the crops grown in the command area grown in the downstream end of the pond, the sizing of the pond will be different. Again the sizing of the pond depends on the choice of the crops in both *kharif* and *rabi* in catchment and command areas. Another factor that decides the sizing of the pond is the water management practice and irrigation methods followed. Since, water is a crucial commodity in the rainfed situation, water saving irrigation (WSI) techniques must be adopted. In one of the WSI techniques, rice may be irrigated when the management allowable deficit (MAD) of soil moisture in the field is 20% below the saturation moisture content during the critical growth stage and other stages of the rice may be purely rainfed. In addition to the above mentioned factors, the sizing of the pond depends on soil and topographic condition of the farm, unlined or lined pond, if lined nature of lining, depth and side slope of the pond.

Panigrahi (2001) reported that 12% of farm area may be diverted for construction of optimum size of pond (plastic lined at both sides and bed of the pond; side slope 1:1 and depth 2.5 m) that can support supplemental irrigation to rice in *kharif* supplied in critical growth stage when the soil moisture is 20% below the saturation moisture content level and pre-sowing irrigation to the mustard in rabi in the farm. A user-friendly software using Visual Basic 6.0 program, has been developed by Roy *et al.* (2009) to simulate the optimum size of the pond in terms of the percentage of the farm area for various rice based cropping systems grown under various irrigation management practices, under both unlined and lined pond systems having different side slopes and depths, for varying field areas and for different types of soil.

One of the most important parameters for determining the optimum size of farm pond is the water balance model of both the crop(s) field and farm pond. Water balance model of crop(s) fields and farm pond helps in assessing the total inflow and outflow to and from the farm pond which will in turn help in computing the size of the pond. Since rice is the predominant crop in eastern India and many other parts in the country in *kharif*/monsoon season, optimum size of rice based cropping system is discussed in this chapter.

7.3 Model Development

7.3.1. Water Balance Model of Rice

The schematic presentation of different water balance model parameters of rice is illustrated in Fig. 7.1. The inflow of the model consists of rainfall R, supplemental irrigation I_s provided to the crop from the pond and contribution of seepage loss of water from the pond to the crop field D_s and upward flux due to capillary rise, U. The outflow consists of surface runoff SR coming from the field and harvested in the pond, seepage and percolation losses S and D, respectively of the crop field and actual evapotranspiration ETcrop. Considering the effective root zone of rice (45 cm) as single layer, the general water balance model of rice under unsaturated phase (no ponding) is:

$$S_{w,i} = S_{w,i-1} + R_i + I_{s,i} - ETcrop_i - SR_i - S_i - D_i + Ds,i + Ui \qquad ...(7.1)$$

where Sw,i is soil water content in the root zone of rice at the end of day i (i is the time index taken as 1 day in the study) and other parameters of the model are defined as above. The units of all the parameters of the model are considered in mm.

Contribution of seepage loss of water from the pond to the field is computed as the volume of seepage loss of the pond water (VS,,i) per its unit wetted surface area. Capillary flux of groundwater may be ignored in the areas where the groundwater table lies more than 2 m below the crop effective root zone. At field scale level, inflow and outflow of groundwater into the rice field have little significance and hence may neglected. Moreover, the horizontal seepage inflow and outflow through the rice field dykes may be assumed constant for the leveled field (Odhiambo and Murty, 1996).

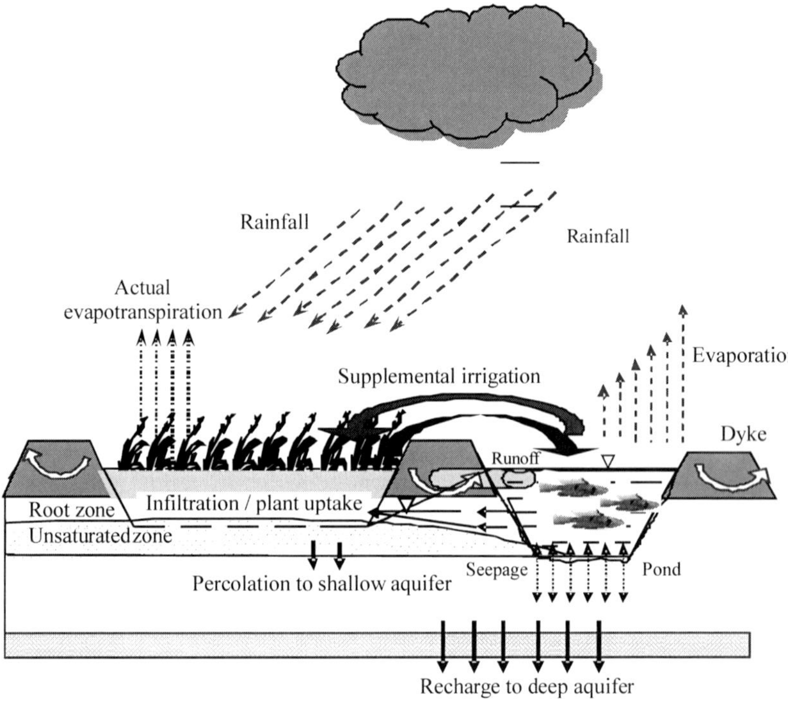

Fig. 7.1. Schematic presentation of water balance parameters of rice field and pond

If soil water content on any day as computed by Eqn (7.1) is more than the saturation water content M_{sat}, then ponding will occur in the rice field. Under the ponding phase, water balance model is:

$$D_{p,i} = D_{p,i-1} + R_i + I_{s,i} - ETcrop_i - SR_i - S_i - Di \quad ...(7.2)$$

where, Dpi is the ponding depth occurred at the end of day i and is computed as:

$$D_{p,i} = S_{w,i} - M_{sat} \quad ...(7.3)$$

Details of computation of the variable/parameters of the water balance model are described below.

7.3.1.1. Actual Evapotranspiration

Actual evapotranspiration in the rice field is:

$$ETcrop_i = k_{si} \times k_{ci} \times ET_{0\,i} \qquad \text{...(7.4)}$$

where: kc is the crop coefficient; ks is the soil water stress factor; i is time index taken as 1 day; and ETo is reference evapotranspiration (mm).

For accurate result, values for ETo should be estimated by FAO Penman-Monteith method (Allen *et al.*, 1998). Values for kc of rice during the crop establishment, crop development, mid season andmaturity stages may be assumed as 1.05, 1.10, 1.10 and 0.95, respectively (Doorenbos and Pruitt, 1977). It is recommended that the locally observed values of kc be taken into account for estimating values of actual evapotranspiration. Value for ks,i is taken as 1.0 under ponding or saturation phase in rice field. However, under unsaturated phase, ks,i is assumed to decrease linearly with the relative available soil water content in the field that is the ratio of soil water content on day i in the 45 cm effective root zone of rice to saturation water content (Panigrahi, 2001; Panigrahi and Panda, 2003).

7.3.1.2. Percolation in Rice Field

Percolation under Ponded Phase:

Assuming the water flow under the ponded/saturated phase as 1D steady state, and the aquifer as isotropic, homogeneous, layered, and isothermal, Laplace equation may be considered for the water flow. Partial differentiation of Laplace equation is:

$$K\left[\frac{\partial^2}{\partial Z^2}\right] + \frac{\partial K}{\partial Z}\frac{\partial \varphi}{\partial Z} = 0 \qquad \text{...(7.5)}$$

where K is the saturated hydraulic conductivity (cm/day); ϕ is the hydraulic head (cm) given as = h+Z, h is piezometric/hydraulic head (cm) and Z is elevation head as height above the reference level (cm). The surface of the water table can be considered as the reference level.

Values of K are determined by field investigation for different soil layers and groundwater table position below the crop effective root zone. A finite difference approach based on a central difference scheme can be used for solution of Eq. (7.5) that predicts the value of the hydraulic head at the required depth. The resultant finite difference approximation of the Eqn. (7.5) is:

$$\varphi(i+1)\left[\frac{K(i)}{(L_1 \times L_2)}+\frac{K(i+1)}{(L_1+L_2)^2}-\frac{K(i-1)}{(L_1+L_2)^2}\right]-\varphi(i)\left[\frac{2K(i)}{(L_1 \times L_2)}\right]+\varphi(i-1)\left[\frac{K(i)}{(L_1 \times L_2)}-\frac{K(i+1)}{(L_1+L_2)^2}+\frac{K(i-1)}{(L_1+L_2)^2}\right]=0 \quad ...(7.6)$$

where L_1 and L_2 are the assumed values of the grid spacing up to the root zone and below root zone to ground water table position, respectively. Substituting in Darcy's equation, the flux in vertical direction/percolation, Di becomes::

$$D=\frac{-K(n_{rz})[\varphi(n_{rz}-1)-\varphi(n_{rz})]}{L_1} \quad ...(7.7)$$

where $K(n_{rz})$ is the hydraulic conductivity at the node on the root zone depth (cm/day), $\phi\ (n_{rz})$ is hydraulic head at the node on the root zone depth (cm), and $\phi\ (n_{rz}-1)$ is hydraulic head at the $(n_{rz}-1)^{th}$ node (cm). Computational procedures of Eq. (7.7) for estimation of percolation are discussed in details by Agrawal *et al.* (2004).

Percolation under Unsaturated Phase

Darcy's law in the discrete form is used to compute flux under unsaturated phase as:

$$D_i=\frac{(H_1-H_2).K_{avg}(\theta)}{T} \quad ...(7.8)$$

where H_1 and H_2 are hydraulic potential and Z_1 and Z_2 are elevation head of the first and second compartments, respectively (cm); T is thickness between first and second compartments (cm) and $K_{avg}^{(\theta)}$ is average hydraulic conductivity of the first and second compartments (cm/day). Detail procedures for estimation of percolation under unsaturated phase employing Eq. (7.8) is available in Agrawal *et al.* (2004).

7.3.1.3. Seepage

For calculation of seepage through the rice basin dykes, Dupuit's approach (Paulo et al., 1995) is followed which is:

$$Q_H = \frac{K_H}{2L}(h_0^2 - h_1^2) \qquad ...(7.9)$$

where Q_H is flow through unit length of basin dyke (m³/day/m), K_H is saturated horizontal hydraulic conductivity (m/day), L = width of basin dyke (m) and h_o and h_1 are hydraulic head inside and outside the basin dyke (m). The determination of K_H, h_o and h_1 along with the necessary values for the study site is reported by Agrawal *et al.* (2004). Total seepage along the dyke is computed by multiplying Q_H with the total length of the basin dykes through which seepage occurs. Dividing the total seepage along the dykes with the areas of the field gives seepage loss in terms of depth unit per day.

7.3.1.4. Irrigation Management Practice

In rainfed farming system, irrigation water is a scarce commodity. Hence, supplemental irrigation from pond is provided only during the critical growth stage (reproductive stage which is assumed as the mid season stage) of rice. In the developed software, though several options on irrigation management practice are allowed, but to simulate the optimum size of pond, four options on when to irrigate rice during the reproductive stage are mentioned. These options are maintaining saturation always and irrigation at management allowable deficit (MAD) of 10, 20 and 30% depletion from saturation. At each irrigation, 5 cm irrigation or the actual available amount in the pond whichever is smaller is applied to rice. Moreover, in irrigated rice, no ponding is allowed in the rice field up to 10 days after germination of seeds to last 10 days before harvest of rice. All the ponded water is drained to the pond constructed at one end of the field. During other periods in rice, excess of 5 cm ponding in field is drained to pond as surface runoff. After harvest of rice, a non rice (dry) crop is sown in the same field.

Unlike rice, dry land crops like groundnut, mustard etc. need less water. They are grown at MAD levels of 20 to 70% depletion of available soil moisture depending on type of the crop and stages of crop growth (Doorenbos and Pruitt, 1977). Options of irrigation timings to dry land crops have been kept in a wide range in the present software that facilitates the farmers to decide the pond size for their preferences.

They are providing pre-sowing irrigation (PSI) to crops for seed germination purpose only; or providing it as PSI and then providing supplemental irrigation during the critical growth stage (crop development and mid season stage) at MAD such that the soil moisture content in the effective root zone of the crop is (i) maintained always at field capacity and (ii) depleted by 20, 30, 40, 50, 60 and 70% available soil moisture. At each irrigation water from pond is provided till the soil moisture comes to field capacity level. Pre-sowing irrigation to dry land crops is required to bring the current available soil moisture content in the effective root zone of the crop to 75% available soil water content (Verma and Sarma, 1990). A soil water balance model of dry land crop is used to compute the soil water status in the effective root zone to decide the timing and irrigation amount to be given from the pond and also to compute the surface runoff from the field.

7.3.1.5. Bare Soil Evaporation

Initially when the seed is sown in the field, the soil moisture is assumed to be at wilting point. During the germination period, there is no crop in the field and bare soil evaporation is computed from reference crop evapotranspiration subjected to rainfall condition as suggested by Jensen et al. (1993). It is assumed that three to five days are needed for germination of seed of both rice and dry land crops sown in the field.

7.3.1.6. Water Balance Model during Turn-in-Period

After harvest of rice, 15 days are required for land preparation called as turn-in-period for sowing of dry land crops. During this period, soil is under bare condition. The water balance model of rice under unsaturated phase is extended till the end of turn-in-period to decide whether pre-sowing irrigation to dry land crop is needed or not. During the said period, the soil water content on any day is:

$$S_{w,i} = S_{w,i-1} + R_i - E_{T,i} - SR_i - S_i - D_i + Ds,i \qquad ...(7.10)$$

where, E_{Ti} is bare soil evaporation during turn-in-period computed as suggested by Sanchez-Cohen *et al.* (1997). If the value of soil water content computed by Eq. (7.10) as above is more than the saturation moisture content, then this excess amount is drained to the pond as surface runoff.

7.3.2. Water Balance Model of Dry Land Crops

The maximum rooting depth of dry land crop is considered as the soil reservoir. The soil reservoir is divided in to 2 layers: (i) an active soil layer in which the roots are present and from which both moisture extraction and drainage occur and (ii) a passive soil layer from which only drainage occurs. During the initial period of crop growth, these two zones exist separately with their relative dimensions being determined by the rate of root growth. When the roots attain maximum depth, the entire root zone is occupied by the active layer. The components of soil water balance model of mustard are shown in Fig. 7.2.

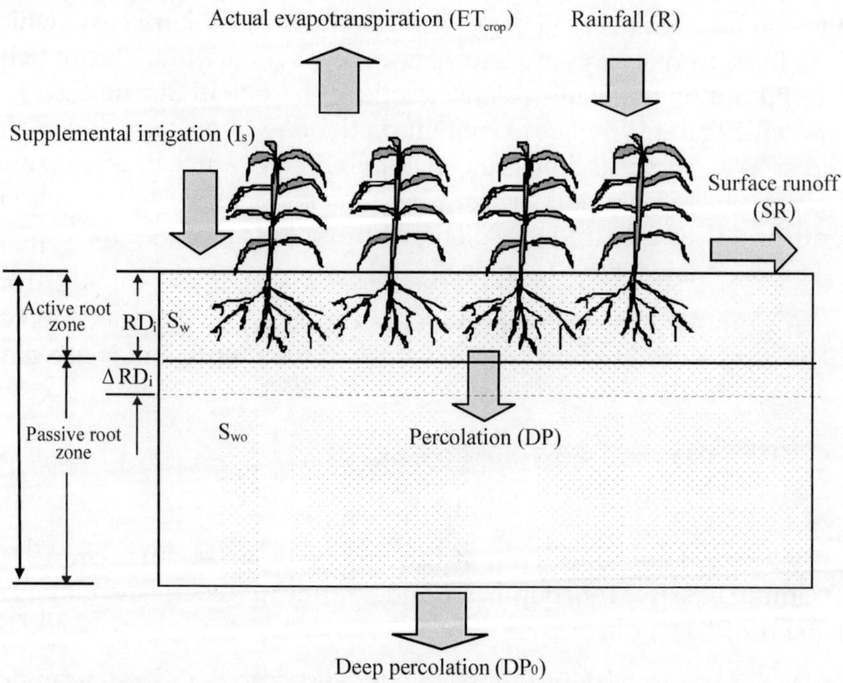

Fig. 7.2. Components of soil water balance model of dry land crops

The daily soil water balance in the active root zone, ignoring upward flux because of capillary rise (Hajilal et al., 1998) is:

$$S_{w,i}.RD_i = S_{w,i-1}.RD_{i-1} + R_i + I_{s,i} + \Delta RD_i.S_{wo,i-1} - DP_i - ET_{crop,i} - SR_i \quad ...(7.11)$$

where, Sw = soil moisture content in the active root zone, mm cm^{-1}; Swo = soil moisture content in the passive root zone, mm cm^{-1}; RD is the root depth, cm; ΔRD=incremental root depth, cm; R=rainfall, mm; SR = surface runoff, mm; Is = supplemental irrigation given to crop, mm; DP = percolation out of active root zone, mm; ETcrop = actual evapotranspiration, mm and i = time index taken as 1 day in the study.

The term (ΔRDi . Swo,i-1) represents the addition of soil moisture content of a soil layer (equivalent to root growth during the period), when the analysis proceeds from (i-1)th to i-th day. However, before this interval is reached, value of soil moisture content in the incremental layer is not known. It is assumed that the soil moisture content of the additional soil layer is same as that of the passive root zone. Values of soil water content in both active and passive root layer at the beginning of the simulation period i.e. germination day of dry land crop, which takes three to five days after sowing of seeds are assumed same. This is found out by the water balance model of bare field during the turn-in period. The soil moisture content on the day of germination of dry land crop as computed by the above mentioned water balance model is given over 15-cm soil depth. The values are converted to per cm depth of soil layer to find the soil moisture content in active and passive layer (Swi-1 and Swo,i-1, respectively).

A sigmoidal root growth model as proposed by Borg and Grimes (1986) can be used to find out the values of root depth and is given as:

$$RD_i = RD_m \left[0.5 + 0.5 \sin \left\{ 3.03 \left(\frac{t}{t_m} \right) - 1.47 \right\} \right] \qquad ...(7.12)$$

where, RD_i = root depth at time t days after sowing, cm; RD_m= maximum possible root depth, cm and t_m=time for the full development of the root zone, days

Maximum root depth for each dry land crop is to be determined from the field experiment data from the root sampling of the crop. In the above mentioned root growth model (Eq. 7.12), the minimum value of RD_i is taken as 15 cm since, soil evaporation can occur from top 15 cm soil layer (Rao, 1987; Hajilal *et al.*, 1998).

Under adequate soil moisture condition, the potential evapotranspiration on i-th day (PET_i) is given as:

$$PET_i = Kc_i . ETo_i \qquad ...(7.13)$$

However, under water stress condition, actual evapotranspiration ($ET_{crop,i}$) is given as:

$$ET_{crop_i} = \frac{(S_{w,i} - WP)}{(1-p).(FC - WP)}.PET_i \text{ if } WP < S_{wi} < (1-p).(FC - WP) \quad ... (7.14)$$

$$ET_{crop_i} = PET_i \text{ if } (S_{w,i} - WP) \geq (1-p).(FC - WP) \quad ... (7.15)$$

where, FC = field capacity, mm cm^{-1}; WP = wilting point, mm cm^{-1} and p = soil moisture depletion factor that is computed as proposed by Doorenbos and Kasam (1979).

Surface runoff generated from the crop field is computed by the US SCS Curve Number method as discussed earlier in Chapter 4. The amount of water percolating below the active root zone is given as:

$$DP_i = [R_i + Is_i - SR_i] - [(FC - S_{wo,i-1}). RD_{i-1} + (FC - S_{wo,i-1}). \Delta RD_i] \quad ...(7.16)$$

Moisture content in the passive root zone changes depending on the value of water percolating below the active root zone as:

$$S_{wo,i} = S_{wo,i-1} \quad \text{if } DP_i \leq 0 \quad ...(7.17)$$

$$S_{wo,i} = S_{wo,i-1} + \frac{DP_i}{(RD_m - RD_i)} \text{ if } DP_i > 0 \quad ...(7.18)$$

If $S_{wo,i}$ calculated by equation (7.17) and (7.18) is more than the soil moisture content at field capacity (FC) in the passive root zone, then there will be deep percolation out of passive root zone. $S_{wo,i}$ then becomes equal to FC of the passive root zone. Deep percolation out of passive root zone (DP_{oi}) is given as:

$$DP_{oi} = DP_i - (FC - S_{wo,i}). \Delta RD_i \quad ... (7.19)$$

Moreover, if the depth of root on any day (RD_i) equals to the maximum root depth (RDm), then there is no passive root layer and then Swo,i becomes equal to zero.

A computer programme in C language is developed to compute the soil moisture content on daily basis in the active and passive root zone layer of the dry land crop by using the soil water balance model. The flow chart for the programme is presented in Fig. 7.3.

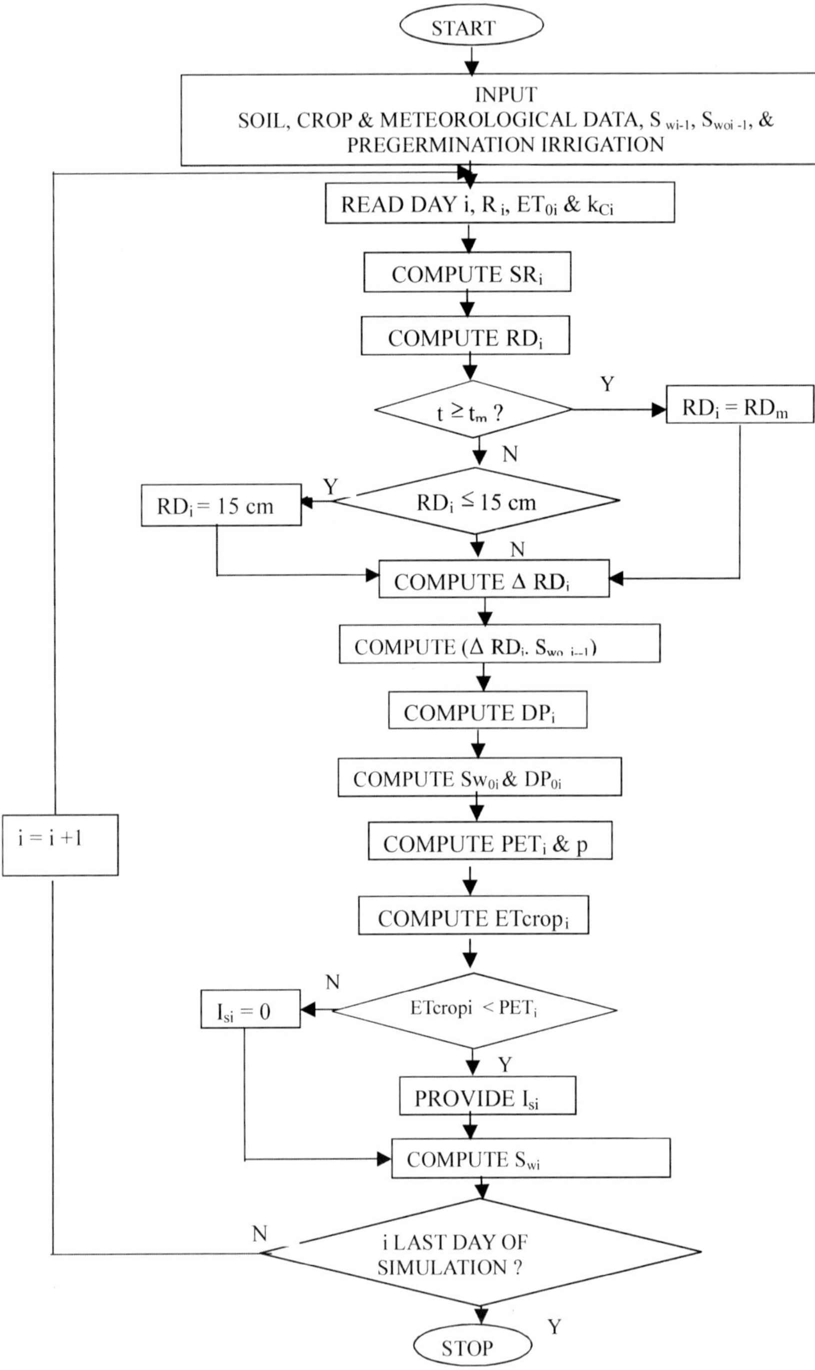

Fig. 7.3. Flow chart for soil water balance model of dry land crop

7.3.3. Water Balance Model of Farm Pond

A farm pond may be located at the end of a field. Runoff from the field (catchment) is harvested in the pond and recycled as supplemental irrigation to the crops grown in the same field. Such a situation is illustrated in Fig. 7.1. Different inflow and out flow components of this farm pond irrigation system are illustrated in the above mentioned figure.

The water balance model of the farm pond is given as:

$$S_R = S_{R-1} + Q_{SRi} + Q_{Ri} - Q_{E1} - Q_{SPi} - Q_{SIi} \qquad \text{... (7.20)}$$

where, S = water storage in the pond, m^3; Q_{SR} = volume of surface runoff coming from the field to the pond, m^3; Q_R = volume of direct rainfall in the pond, m^3; Q_E = volume of water lost as evaporation from the pond, m^3; Q_{SI} = volume of water used as supplemental irrigation in the cropped field, m^3; Q_{SI} = volume of water lost as seepage from the pond, m^3 and i = time index taken as 1 day in the study.

In the above water balance model, values of Q_{SR} is given as:

$$Q_{SRi} = SR_i.A_{field} \qquad \text{...(7.21)}$$

where, A_{field} = field area which is the total farm/field area minus the area lost to construction of the farm pond. If a farm area is 1000 m^2 and in it a farm pond of area 100 m^2 is constructed, then the field area is 900 m^2.

Surface runoff from the catchment depends on the crop and water management practice of the crop in the catchment. For rice grown in the catchment, some standing water is allowed in the field whereas for dryland crops, no water is allowed in the field. For paddy, excess of ponding than the management allowable ponding depth (normally 5-10 cm) is the surface runoff stored in the pond. For dry land crop, US SCS Curve Number method (discussed in Chapter 4) may be used for computation of surface runoff.

Volume of direct conservation of rainfall in the pond is:

$$Q_{Ri} = R_i.A_{OFR} \qquad \text{...(7.22)}$$

where R_i is the rainfall contributed to the pond with area A_{OFR}.

Volume of water used as supplemental irrigation (Q_{SIi}) to the crop from the pond on any day can be found out by measuring the discharge of the pump used for irrigation and the time of irrigation by the pump. Volume of evaporation loss in the pond on any day depend on the value of the pond evaporation of that day and the water spread area at the top surface of the pond on that day (Awsi). Computation of pond evaporation is discussed in Chapter 9. Volume of seepage loss in the farm pond on any day depends on the value of the seepage loss rate of that day and the wetted surface area of the pond on that day (Awsi). Computation of seepage loss in the farm pond is also discussed in Chapter 9. It is to be noted that farm ponds constructed in the areas with ground water table very near to the ground level may have reverse flow instead of seepage. This may be designated as ground water contribution. Depending on the hydraulic gradient between the ground water table position and the level of stored water in the pond, ground water contribution varies (Roy, 2006).

7.4 Simulation Study

The simulation period for computation of pond size for a year should start from the onset of monsoon season of that year when the rice is normally dry-seeded in the field till the harvest day of the crop which is generally harvested before the withdrawal of monsoon. The onset and withdrawal of monsoon for the eastern regions of the country has been predicted by Panigrahi (2001) which are June 16th and October 3rd, respectively. Thus, the monsoon is effective in the region for 110 days. Hence, it is recommended that a short duration rice of about 110 days would be appropriate that could effectively utilize the maximum rainwater of the monsoon season. After the harvest of rice, dry land crop is to be grown in the same field where rice is grown. The durations of crop establishment, crop development, mid season and maturity stages of the rice from the day of germination till harvest for a short duration rice (101 days) are taken as 25, 20, 30, and 26 days, respectively. The pond size is determined for various rice based cropping systems. However, to demonstrate the result, the software developed is applied to rice-mustard and rice-groundnut cropping systems. The crop input parameters of rice, groundnut and mustard of the study area are to be fed to the model.

In the simulation study, in order to verify the simulated results of the pond sizes, field areas ranging from 1000 to 10000 m^2 are assumed. In the simulation, it is assumed that the entire crop field from which the runoff was drained out and harvested in the pond was covered with rice in monsoon season and dry land crop in winter and was provided with irrigation from the pond. The dimensions of the dykes around the pond are assumed as: 30 cm berm width, 30 cm each of top width and height of bund, and 1:1 side slope. Software developed and a program written in Visual Basic 6.0 to compute the size of the farm pond is presented in Fig. 7.4.

7.4.1. Description of Software

For simulating the pond size, the model needs initial value of the soil water content when the rice seed is sown in the dry field that is assumed as the soil water content at wilting point, field area in which the pond is to be constructed, initial pond size that is assumed as 5% of the field area, dimensions of the pond, soil, crop and different meteorological data of the study site on daily basis, initial pond storage, irrigation management practice of the crops during different growth stages etc. On the basis of the above data, water balance model of the rice field as described by Eqns. (7.1) to (7.9) simulates the values of actual evapotranspiration, seepage, percolation, surface runoff generated from the field and supplemental irrigation requirement as per the pre-assigned irrigation management practice on daily basis. At the same time, the pond water balance model (Eqn. 7.20) is run to predict the daily pond storage. The water balance model of the dry land crop is also run simultaneously to decide the irrigation scheduling as per the pre-assigned irrigation management practice of the dry land crop.

The simulation continues from the onset day of the monsoon when the rice seed is sown as dry seeded until the harvest of the dry land crops. During the simulation period, if the pond storage is found to be insufficient to meet the pre-assigned irrigation demands of the crops as proposed earlier (section Irrigation Management Practice), then the pond is considered as under-designed. Under such a situation, the pond size is increased by 0.1% over the initial assumed size of 5%of the farm area and the simulation is again continued for the same year from the onset day of monsoon. The simulated pond sizes vary from year to year due to the stochastic nature of the variables like rainfall, runoff, evapotranspiration and related water balance parameters. Normal distribution or any other suitable distribution may be fitted to the data series containing different pond sizes for each type and condition and the expected pond sizes for different probability of exceedences (PE) should be predicted by using the SMADA software (Wanielista *et al.,* 1997).

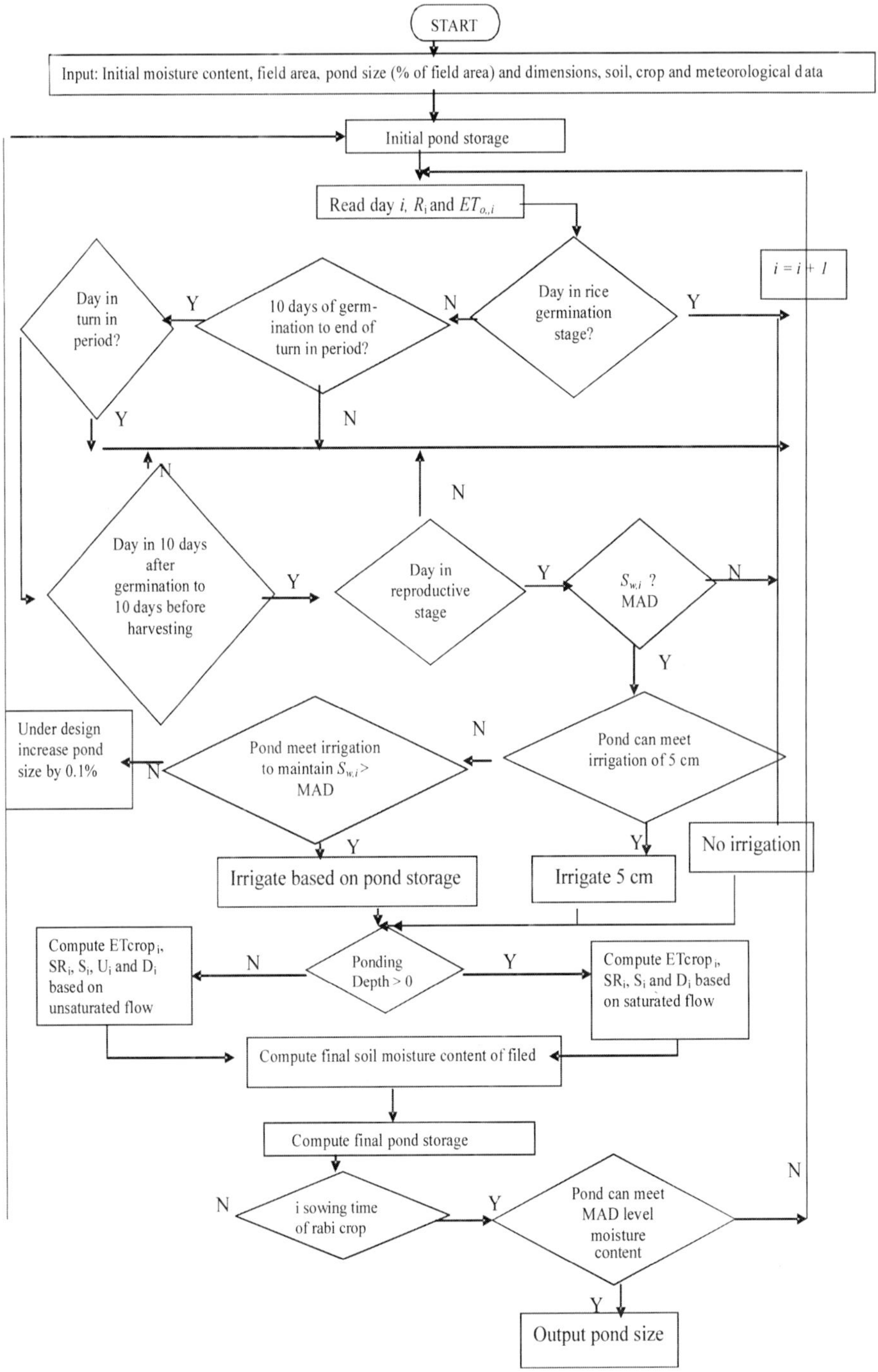

Fig. 7.4. Flow chart for computation of pond size

7.4.2. Main Window

Various inputs to this window are: (i) side slope of the pond (ii) types of lining materials used in the pond (iii) field areas in which pond is constructed (it includes both the catchment and command areas) (iv) soil types of pond site (v) pond dimensions (vi) initial pond size and (vii) irrigation management practices of the crops grown in different seasons. In this window, user can specify the depth of pond. In that case the software will simulate the size of pond only for that specified depth. One combo box is provided to select the soil type. As user selects the soil type, the software will read the required data from a saved file. If user is unable to find the soil type of his choice, he can add a new soil type by incorporating required soil physio-chemical information. Next time when he runs the software, the soil type will appear in the combo box.

The software is flexible for various crops and cropping systems. User can select the preferred crops from the combo boxes. Once the crop is selected, the software will automatically read necessary information about the crop from a data file. If the user cannot find the desired crop in the combo boxes, option is provided to add crops of his choice. The window for adding a new crop to the software is shown in Fig. 7.5. The user has to enter various crop related inputs from this figure and click the save button in the crop file. Next time, the said crop will appear in the combo box when the user runs the software by using the crop information to simulate the pond size. Various inputs to this window are: (i) name of the crop(s) (ii) effective root zone depth of the crop(s) (iii) crop(s) growth season (iv) number of crop(s) growth stages (v) duration of each growth stage with corresponding crop(s) coefficient value (vi) critical crop(s) growth stage requiring irrigation and (vii) crop(s) stress factor.

7.4.3. Soil Window

The software can be used for various soil types as selected by the user. Accordingly, the software will read necessary data from a saved data file. Similar to crop option, user can add a custom soil type by entering some basic data needed by the software and can save it in the combo box for further use. User has to enter various soil related data as: (i) water holding capacity (ii) soil moisture content at field capacity and wilting point (iii) saturation moisture content (iv) saturated hydraulic conductivity (v) residual soil moisture content (vi) texture specific empirical constants (vii) bubbling pressure and (viii) permeability at different depths in the soil. In addition, provisions for windows for meteorological data including rainfall, window for entering crop evapotranspiration data etc. are made in the software while simulating the pond sizes.

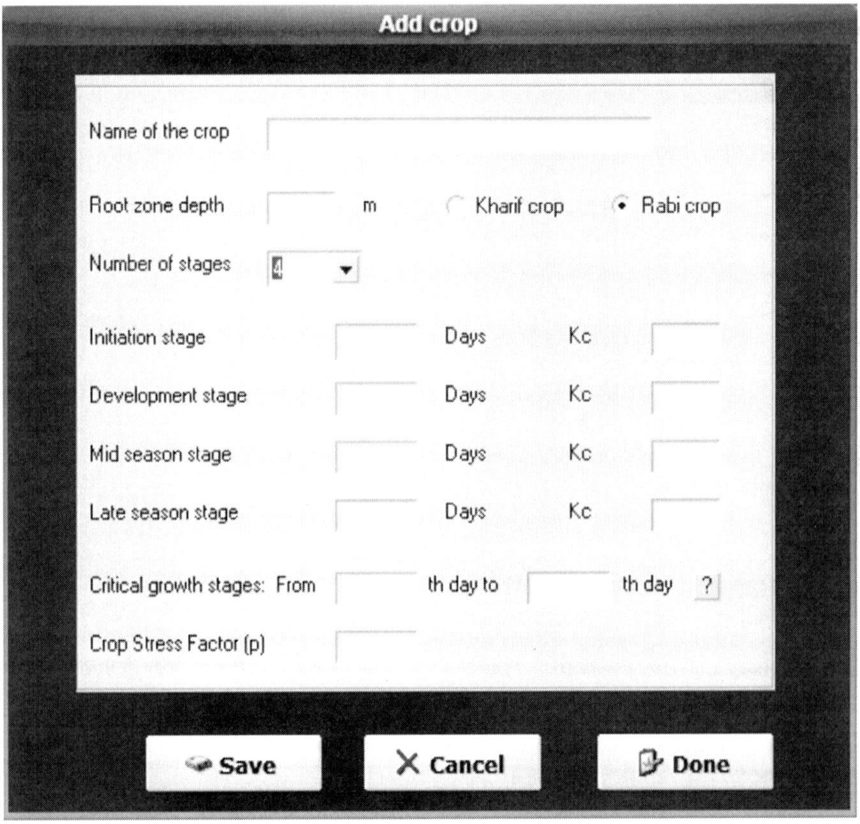

Fig. 7.5. The window for adding a new crop

7.4.4. Window for Displaying Results

The window for displaying results is shown in Fig. 7.6. If the software is run without specifying the pond depth, it will start simulating the pond sizes from initial depth with an increment of 0.05 m. After the simulation is over, the "show result" button is to be pressed which will display the pond sizes for each depth on the picture box. Among these various combinations, the software will select the minimum pond size corresponding to a particular depth. It will do the job by hitting of the "calculate" button. However, if the specified depth in the input box is entered, then the software will calculate the optimum pond size along with various dimensions of the pond. A schematic diagram will also be displayed on the right hand side picture box of the window.

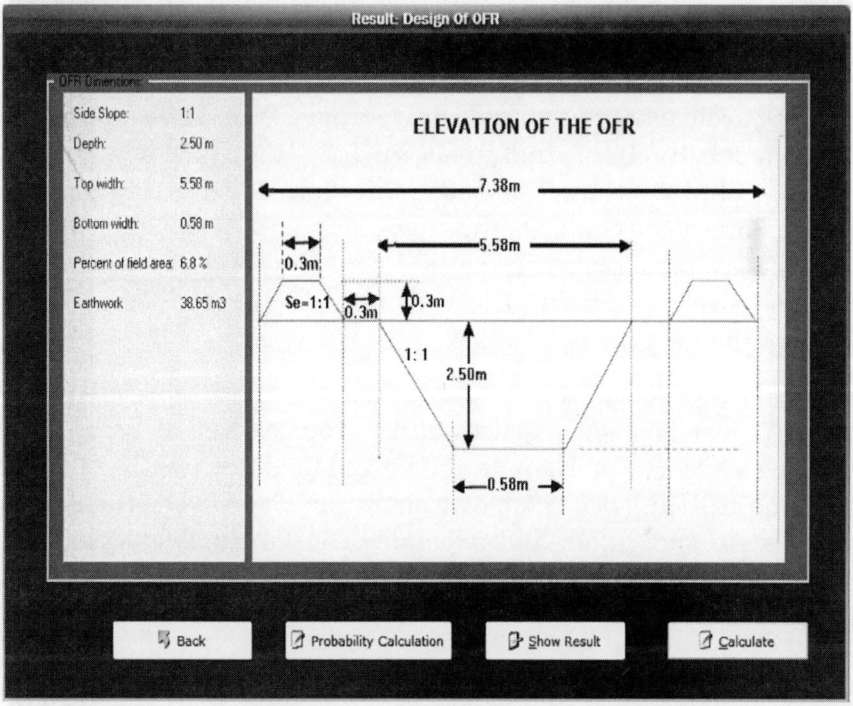

Fig. 7.6. Window to display simulation size of pond

7.5 Simulation of Pond Size

Pond sizes (percent of field area) are simulated by the software for 30 years of hydrological data for various combinations of cropping systems, irrigation management practices, soil, pond lining conditions, side slope and depth of ponds and farm sizes. The results of the pond sizes obtained for two main cropping systems i.e. rice-mustard and rice-groundnut are presented here. The pond sizes are simulated for (i) field areas ranging from 1000 to 10000 m^2 (ii) two side slopes (horizontal : vertical) of 1:1 and 1.5:1 (depth considered 2.5 m for both the side slopes with plastic lining by 250 micron films) (iii) irrigation management practices for *kharif* crops being at two levels i.e. irrigation at MAD value of 10 and 20% and for *rabi* crops at 40 and 50% depletion of available soil moisture. The optimal sizes obtained for rice mustard and rice groundnut cropping systems for various field areas, pond side slopes of 1:1 and 1.5:1 (fixed depth of 2.5 m) and different irrigation management practices for plastic lined case with medium land topography are presented in Tables 7.1 and 7.2, respectively.

The optimal pond sizes decrease as the field area increase for both the cropping systems. When the field areas increase from 1000 to 10000 m^2, the optimal pond sizes for rice-mustard and rice-groundnut cropping system are found to decrease from 13.1 to 11.5% and 13.8 to 12.3%, respectively of field areas for irrigation schedules of 10% depletion of MAD in *kharif* and 40% depletion of MAD in *rabi* crops at 1:1 side slope of the pond. For the same cropping systems and irrigation management practices when the side slopes of pond increase to 1.5:1, the pond sizes are observed to vary from 14.0 to 12.4% and 14.9 to 13.3% of the field areas.

It is observed that the sizes of ponds are dependent on the irrigation management practices of the crops as well as types of the cropping systems. For example, for a given field size (say 10000 m^2), if irrigation to *kharif* rice changes from 10 to 20% MAD (irrigation to *rabi* mustard and ground nut remaining constant at 40% depletion of available soil moisture), then pond sizes decrease from 11.5 to 11.0% and 12.3 to 11.7% of field area for rice-mustard and rice-groundnut cropping systems, respectively. Rice-groundnut cropping system is found to require large size ponds than the rice-mustard cropping system.

The software has been tested with two cropping systems i.e. rice-mustard and rice-groundnut to simulate the optimal pond size for lined condition for 1:1 and 1.5:1 side slopes of pond with 2.5 m depth in each case. The rice-groundnut cropping system is found to require large size ponds than the rice-mustard cropping system. The reason of requirement of larger pond size for rice -ground nut cropping system than the rice – mustard one is due to the higher irrigation requirement of ground nut crop than the mustard. Mustard needs hardly 2 to 3 irrigations where as ground nut requires 5-6 irrigations. Because of higher irrigation requirement for rice- ground nut cropping systems, larger storage structure of the pond is required which necessitates higher pond size to be constructed.

Table 7.1. Optimum size of pond for rice mustard cropping pattern

Field area, m^2	Optimum pond size, % of field area							
	Pond side slope, 1:1				Pond side slope 1.5:1			
	Irrigation management practices				Irrigation management practices			
	I_1	I_2	I_3	I_4	I_1	I_2	I_3	I_4
1000	13.1	12.9	12.6	12.4	14.0	13.8	13.6	13.3
2000	12.9	12.7	12.4	12.2	13.8	13.6	13.5	13.1
3000	12.7	12.5	12.3	12.1	13.6	13.5	13.4	13.0
4000	12.6	12.4	12.2	11.9	13.4	13.2	13.0	12.8
5000	12.4	12.3	12.0	11.8	13.3	13.1	12.8	12.6
6000	12.2	12.1	11.8	11.6	13.1	13.0	12.7	12.5
7000	12.1	12.0	11.7	11.4	13.0	12.8	12.5	12.3
8000	11.9	11.9	11.5	11.1	12.8	12.7	1′2.4	12.0
9000	11.6	11.6	11.2	11.0	12.6	12.5	12.3	11.7
10000	11.5	11.3	11.0	10.8	12.4	12.2	12.0	11.5

N.B. I_1 = 10% depletion of management allowable deficit (MAD) in *kharif* rice and 40% depletion of MAD in *rabi* crops; I_2 = 10% depletion of management allowable deficit (MAD) in *kharif* rice and 50% depletion of MAD in *rabi* crops; I_3 = 20% depletion of management allowable deficit (MAD) in *kharif* rice and 40% depletion of MAD in *rabi* crops and I_1 = 20% depletion of management allowable deficit (MAD) in *kharif* rice and 50% depletion of MAD in *rabi* crops.

It is to be remembered that the user-friendly software developed by using the Visual Basic 6.0 program is used to find out the optimal size of the pond in terms of percentage of field area (here in simply called as the pond sizes) by simulating the water balance model parameters of the crop field and the pond. The menu driven system of the software is flexible enough to simulate the pond sizes for various combinations of the pond geometry, field sizes, and the cropping systems. The user has to specify the crops to be grown in the fields, irrigation management practices of the crops, types of pond i.e. lined or unlined, side slope and depth of the pond and field sizes. The software is then run and the pond size is finally simulated for each case.

Table 7.2. Optimum size of pond for rice groundnut cropping pattern

Field area, m^2	Optimum pond size, % of field area							
	Pond side slope, 1:1				Pond side slope 1.5:1			
	Irrigation management practices				Irrigation management practices			
	I_1	I_2	I_3	I_4	I_1	I_2	I_3	I_4
1000	13.8	13.6	13.2	13.0	14.9	14.7	14.5	14.2
2000	13.7	13.4	13.1	13.0	14.7	14.5	14.2	14.0
3000	13.5	13.3	13.0	12.8	14.5	14.3	14.1	13.9
4000	13.3	13.2	12.8	12.6	14.3	14.1	14.0	13.7
5000	13.0	13.0	12.6	12.4	14.1	14.0	13.8	13.5
6000	13.0	12.8	12.5	12.3	14.0	13.8	13.6	13.3
7000	12.8	12.6	12.4	12.1	13.8	13.6	13.4	13.1
8000	12.6	12.4	12.2	12.0	13.6	13.4	13.2	12.8
9000	12.4	12.2	12.0	11.7	13.5	13.2	13.1	12.6
10000	12.3	12.0	11.7	11.5	13.3	13.0	12.8	12.5

N.B. Notations of symbols of I_1, I_2, I_3 and I_4 are same as mentioned in Table 7.1.

Question Banks

Q1. What do you mean by under-designed and over-designed farm ponds?

Q2. What is an optimum size farm pond? Why it is important to construct optimum size farm pond in field?

Q3. What are the factors on which the size of a farm pond depends on?

Q4. How does the size of a farm pond get affected with the crop and cropping systems?

Q5. Irrigation management practice has a significant influence on deciding the size of a farm pond. Justify the statement.

Q6. Write the general water balance model of rice under both ponded and unsaturated phase and explain the various terms of the models.

Q7. Draw a well labeled diagram of schematic presentation of water balance model of rice field having a pond constructed at one corner.

Q8. How is the actual evapotranspiration of rice computed? How does the actual evapotranspiration get affected as the ponded water in the field vanishes and unsaturated phase occurs?

Q9. Differentiate between seepage and percolation? How does the water depth in the field influence both the parameters?

Q10. How can you estimate the percolation in rice field under both ponded and unsaturated condition?

Q11. Write the Dupuit's formula to predict seepage loss through the rice basin dykes and describe how this formula is used to predict the seepage loss.

Q12. What do you mean by turn-in period? How can you find out the soil moisture content in the field during this period?

Q13. Write the general water balance model of dry land crop field and explain the various parameters of the model.

Q14. Draw a schematic and labeled diagram of soil water balance model of dry land crop field?

Q15. Draw a flow chart for prediction of soil moisture content in the crop field under unsaturated condition.

Q16. Write the water balance model of farm pond and describe how the different parameters of the model are computed.

Q17. Draw the flow chart for simulation of farm pond size and describe in brief how it is helpful to simulate the size of farm pond.

Q18. What do you mean by user-friendly software to predict the pond size in a field? How does the crop and soil window help in simulation of pond size?

Q19. Does the crop and cropping systems in a field affect the pond size? Which type of the cropping system in respect to rice-mustard and rice-ground nut will require bigger size of farm pond and explain why?

Q20. How does the area of the farm affect the size of the farm pond?

Q21. If the side slopes of a farm pond changes from 1:1 to 1.5:1, then how will it affect the size of farm pond.

Q22. If the irrigation management practices to a crop changes from 20 to 40 percent depletion level of management allowable deficit, then how does the farm pond size get changed.

References

Agrawal, M.K., Panda, S.N., Panigrahi, B., 2004. Modeling water balance parameters for rainfed rice. J. Irrg. Draing. Eng., ASCE, 130 (2), 129-139.

Allen, R.G., Pereira , L.S., Raes, D. and Smith, M. 1998. Crop Evapotranspiration: Guidelines for Computing Water Requirements. Irrigation and Drainage Paper 56, FAO, Rome, pp 300.

Borg, H and Grimes, D.W. 1986. Depth development of roots with time, an empirical description. Trans. ASAE, Vol. 29(1): 194-197.

Doorenbos, J. and Kassam, A.H. 1979. Yield Response to Water. Irrigation and Drainage Paper 33, FAO, Rome.

Doorenbos, J. and Pruitt, W.O. 1977. Guidelines for Predicting Crop water Requirements. Irrigation and Drainage Paper 24, FAO, Rome

Hajilal, M.S., Rao, N.H. and Sarma, P.B.S. 1998. Planning intraseasonal water requirements in irrigation projects. Agric. Water Manage., Vol. 37: 163-182.

Jensen, J.R., Mannan, S.M.A. and Uddin, S.M.N. 1993. Irrigation equirement of transplanted monsoon rice in Bangladesh. Agricultural Water Management, Vol. 23(3):199-212.

Odhiambo, L.O. and Murty, V.V.N. 1996. Modeling water balance components in relation to field layouts in low land paddy fields. I. model development, Agricultural Water Management, Vol. 30(2): 185-199.

Panigrahi, B. 2001. Water Balance Simulation for Optimum Design of On-Farm Reservoir in Rainfed Farming System. Ph.D. Thesis, Indian Institute of Technology, Kharagpur.

Panigrahi, B., Panda, S. N., 2003. Optimal sizing of on-farm reservoirs for supplemental irrigation. J. Irrig. Drain. Eng., ASCE, 129(2), 117-128.

Paulo, A.M., Pereira, L.A., Teixeira, J.L., Pereira, L.S., 1995. Modeling paddy rice irrigation. Crop water simulation models in practice, L.S. Pereira, B.J. van Brock, P. Kabat and R.G. Allen eds., Wageningen, The Netherlands, 287-302.

Rao, N.H. 1987. Field test of a simple soil water balance model for irrigated areas. J. Hydrol., Vol. 91: 179-186.

Roy, D., 2006. Development of a software for the design of on-farm reservoir under rainfed farming system. Unpublished Master Thesis, Agricultural and Food Engineering Department, Indian Institute of Technology, Kharagpur, p 85.

Roy, D; Panda, S.N. and Panigrahi, B. Water Balance Simulation Model for Optimal Sizing of On-Farm Reservoir in Rainfed Farming System. Computers and Electronics in Agriculture, (Elsevier Publication), Vol. 65: 114-124.

Sanchez-Cohen, I., Lopes V. L., Slack, D. C., Fagel, M. M., 1997. Water balance model for small-scale water harvesting systems. J. Irrig. Drain. Eng., ASCE, 123 (2), 123-128.

Verma, H. N., Sarma, P. B. S., 1990. Design of storage tanks for water harvesting in rainfed areas. Agric. Water Manage. 18, 195-207.

Wanielista, M., Kersten, R., Eaglin, R., 1997. Hydrology, Water Quantity and Quality Control. John Wiley and Sons, Inc., 567 pp.

Chapter 8

Construction of Farm Pond

8.1 Introduction

After deciding the optimum size of farm pond and its dimensions, the next step begins the construction of farm pond. Further the site selection and location of the pond in the selected site to be finished before the construction of pond starts. The selection of site and location of farm pond areas have been discussed in Chapter 3. In this chapter, discussion on methodologies for construction of farm pond including spillways, inlet and outlet structures, design of grassed waterways and embankment have been presented. As a guiding principle, the construction of farm pond involves the following steps:

8.2 Site Clearing

The area where the pond is to be constructed has to be inspected first. The site where pond has to be dug out should be cleared to an extent of about 10 m from all sides of top dimensions of the pond. All bushes, shrubs, stumps and other unwanted materials should be dug out and removed. If there are roots of plants, they should be uprooted.

The land may be ploughed and harrowed, if necessary, to get a more or less even topography.

8.3 Demarcation of Pond Area

The pond area is demarcated by driving pegs to indicate the four corners. The four sides of the pond area should be at right angles to each other if a square or rectangular shaped pond has to be dug out. This can be achieved by 3-4-5 rules of surveying. According to this rule, a corner point of a site is chosen and then a rope of length 3 m is laid on one side of the pond area. From the same point another rope of length 4 m is laid on the adjacent side of the same chosen pond area. The two terminal ends of the ropes of 3 and 4 m lengths should be 5 m apart. If not, then one of the rope laid on one side has to be adjusted till the distance between the terminal ends of the ropes remain 5 m apart. This ensures the two sides of the pond area to remain at right angles. The other corner points of the pond area are selected in the similar ways.

8.4 Establishment of Reference Level

Spot levels at the four corners and at the end point of the demarcated farm pond site are taken with reference to a near by temporary bench mark. The average of the levels is transferred on to a permanent or semi permanent object at an approximate distance of about 15-20 m from the pond site (Fig. 8.1). The reference level helps in checking out the required depth of the pond when the pond excavation is in progress.

8.5 Stepping Method of Construction

It is difficult to dig a pond exactly in the trapezoidal shape with uniform side slope. Hence, a segment wise construction otherwise called as stepping method of construction is adopted during the time of actual excavation (Belgaumi *et al.*, 1997). These steps help in digging and lifting out the soil of the pond. Further, if the pond is required to be lined by plastic sheets, then it becomes easy to spread the plastic sheets along these steps and then cover the sheet with soil till a regular and uniform side slope is obtained. In case of unlined pond, these steps are subsequently eased out to get the required shape and designed side slopes.

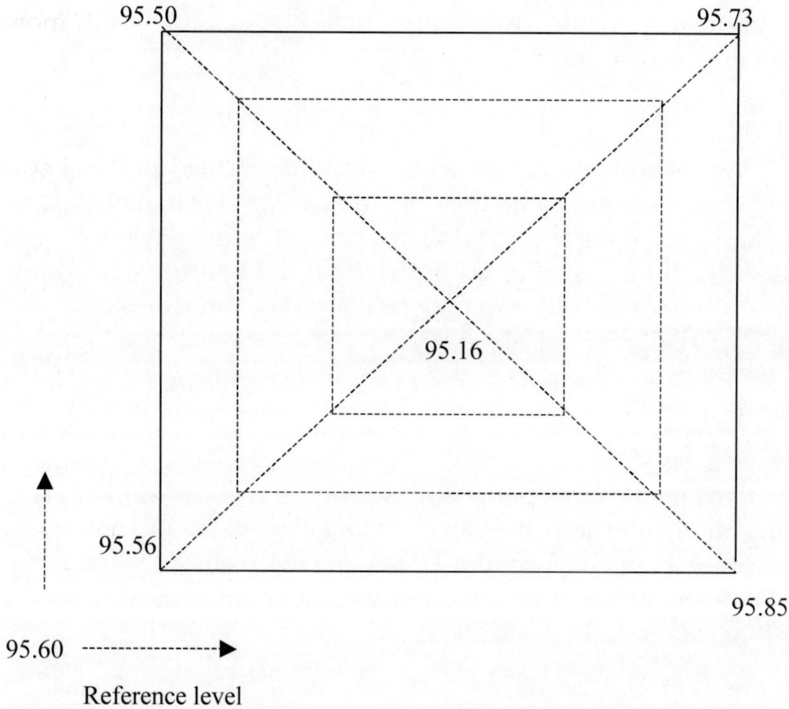

Fig. 8.1. Demarcation of pond area and establishment of reference level

8.6 Excavated Spoil

Since, considerable quantity of soil is obtained while digging out the pond, it is essential to dispose off the same in a proper manner so that it does not create hindrance in movement. The dug out soil, especially the top soil is fertile and should be uniformly spread out over the catchment and command areas of the pond to fill out the depressions and gullies. This ensures a good leveling of the farm pond adjacent areas. Some of the soil can be used to create embankment or dykes of the pond for an embankment type pond. The embankment of the pond provides stability to the pond, increases the storage capacity of the farm pond and at the same time helps in growing plants like papaya, banana, guava, lemon etc.

8.7 Shoulder Bund and Toe Drain

The rainwater falling on the spoil bank and the berm is likely to enter into the pond and create rills all-around. To prevent such rilling, shoulder bund with a small toe drain which would run all along the spoil bank may be provided.

8.8 Silt Trap

A silt trap of suitable dimensions should be formed in the water course just near the entrance to the inlet in order to check the bed load entering the pond along with runoff from the catchment. The length of the silt trap should be slightly grater than the width of the water course. The depth of the silt trap may be 0.5 -1.0 m and side slope may be 1:1 or 1.5:1 (Belgaumi et al., 1997).

8.9 Design of Spillway

8.9.1 Inlet for the Pond

The inlet for the farm pond is designed as a chute spillway for conducting the runoff into the pond. The entry section of the pond may be designed as a rectangular broad crested weir. The discharge through the rectangular broad crested weir without end contraction is given (Sharma and Sharma, 2002) as:

$$Q = 1.84\ L\ H^{1.5} \qquad \text{... (8.1)}$$

where, Q = Discharge rate, m^3/sec; L = Length of the crest, m and H = Depth of flow over the crest, m which is nearly taken as 0.3 m for small size pond.

If the weir has end contraction at both the sides, then the discharge formula is given (Sharma and Sharma, 2002) as:

$$Q = 1.84\ (L-0.2\ H)\ H^{1.5} \qquad \text{... (8.2)}$$

In stead of a broad crested weir, if a trapezoidal weir is used as the inlet section, then the discharge through it is given as:

$$Q = 1.86\ L\ H^{3/2} \qquad \text{... (8.3)}$$

where, Q = discharge, m^3/sec; H = height of water on the crest, m and L = effective length of the crest, m which is given as:

$$L = (L1 + L2)/2 \qquad \text{... (8.4)}$$

The value of the length of crest of the inlet section with a rectangular broad crested weir type spillway is computed by knowing the peak rate of discharge (runoff) coming from the catchment to the pond. After having computed the length of the crest of the weir and height of crest, provide an allowance of 20% extra height for the free board. The peak discharge rate is given either by rational formula (Eq. 4.8) or by USDA SCS Curve Number method (Eq. 4.23) which has already been discussed in computation of water supply to the farm pond in the Chapter 4. Table 8.1 gives the length of the crest of a broad crested weir with and without end contraction for various peak discharges passing over the weir for 0.3 m height of water over the crest.

8.9.2 Outlet for the Pond

Mechanical spillway and emergency spillway are the most important outlet structures in pond design. They are used for disposal of excess water from pond storage. The kind of spillway to be used depends upon the size of watershed contributing runoff to the pond. For a pond having watershed area less than 4 ha, the vegetative spillway can be used. A combination of vegetative and mechanical spillway is provided in the watershed area within 4-12 ha. If the area of watershed is more than 12 ha, care should be taken for the construction of mechanical spillway including its stability (Coche *et al.*, 1995). Mechanical spillway also depends on the drop height. This type of spillway is also called surplus weir and is constructed on a firm foundation.

Emergency spillway is provided at one end of the embankment and is functioned to protect the embankment from overtopping due to unexpected increase in inflow into the pond. Due to high intensity of rainfall occurring for a long period in a catchment or due to unexpected flow from the upper side of the catchment rushing onto the farm pond The bottom of the emergency spillway should be fixed at the maximum flood level (MFL) for the designed runoff and is given by Eq. (8.5) as:

$$Q = CLH^m \qquad \ldots (8.5)$$

where Q = designed peak runoff rate, m^3/sec; L = length of the weir crest, m; H = head of water over the crest, m; C = coefficient of discharge and m = an exponent constant

The values of C and m vary depending upon the type of weir (Senapati, 2008) and are given in Table 8.2.

Table 8.1. Computation of length of crest of a rectangular broad crested weir for various peek discharges (height of water over the crest is taken as constant i.e. 0.3 m)

S. No.	Peak discharge, m^3/sec	Length of crest, m	
		Without end contraction	With end contraction
1	0.1	0.33	0.39
2	0.2	0.66	0.72
3	0.3	0.99	1.05
4	0.4	1.32	1.38
5	0.5	1.65	1.71
6	0.6	1.98	2.04
7	0.7	2.32	2.38
8	0.8	2.65	2.71
9	0.9	2.98	3.03
10	1.0	3.31	3.36
11	1.1	3.64	3.70
12	1.2	3.97	4.02
13	1.3	4.30	4.35
14	1.4	4.63	4.69
15	1.5	4.96	5.02
16	1.6	5.29	5.35
17	1.7	5.62	5.68
18	1.8	5.96	6.01
19	1.9	6.28	6.34
20	2.0	6.69	6.68
21	2.1	6.94	7.00
22	2.2	7.27	7.33
23	2.3	7.60	7.67
24	2.4	7.93	8.00
25	2.5	8.27	8.33

Table 8.12. Values of C and m for different weir types

Type of weir	C	m
Rectangular	1. 84	1.5
Triangular	1. 38	2.48
Trapezoidal	1. 86	1.5

Head of water over the emergency spillway should not be more than 30 cm. Side slope of the trapezoidal emergency spillway is recommended as 2:1 (V:H).

8.9.3. Procedure to Install Weir in Inlet and Outlet of Pond

The following instructions must be followed before installing a weir in the inlet and outlet section of the farm pond (Lenka, 2001).

- The channel section must be straight. The straight section must be at least 10 times the length of weir crest in the upstream.
- Place the weir at right angle to the direction of water flow. Face of the weir must be perpendicular, the weir crest should be straight and level. It should not lean to either to the upstream or down stream side.
- There should not have any obstruction in the upstream side.
- Crest and sides of the weir notch should not be more than 3 mm in thickness. The notch should be of regular shape and its edges must be rigid and straight.
- The flow velocity in the upstream should be uniform. The velocity of approaching flow should not be more than 15 cm/sec. If the flow velocity is non-uniform, then baffle structures may be put in the weir pond t o reduce the flow velocity.
- The centre line of the weir should be parallel to the direction of flow.
- The crest of the weir should be level so that water passing over it will be of same depth at all along the crest.
- The weir crest should be above the bottom of the approach channel at least twice the depth of water flowing over the weir. A minimum distance of about twice the depth of flow should be maintained between the end of the weir crest and the bank of the up-stream channel.
- The crest of the weir is placed high enough so that water will fall freely below the weir: leaving the air space under the over-falling sheet of water. The depth of water flowing over the rectangular weir should not be less than 5 cm and not more than 2/3 of crest width.

- The head-measuring scale should be so placed in the up-stream that it is located at a distance of about 4 times the approximate head: far enough to one side to that it will be completely in still water. The zero of the scale must be exactly on the same level as the bottom of the crest level of the weirs or at the apex of the "V" notch weir.

8.10 Grassed Waterways for Safe Disposal of Water

Excess water from the farm pond should be evacuated through outlet of the pond which is finally drained out by grassed waterways without causing soil erosion or flooding. Thus, the grassed waterways are primarily used for safe disposal of runoff water and/or excess stored water of the farm pond. They may be natural or constructed channels having suitable vegetation linings. They may also be used as inlet in the farm ponds to allow the runoff to get filtered and passed from the catchment to the adjacent farm ponds. The water of the catchment passing in these inlets is relatively sediment free. Thus, they also help in reducing the sedimentation of ponds and thus help in increasing the life of the pond. However, during the process of filtration, they themselves loose their effectiveness when too much sediment builds up in the waterways. These waterways are laid in soil having slope up to 5% and having good vegetation linings. The waterways may be trapezoidal, rectangular or triangular in shape. The installation of the waterways is to be done carefully so that the water flowing in the channels does not create any gully.

8.10.1. Vegetation Establishment

The most important time in installing the grassed waterways is when vegetation is being established. Before establishing the vegetations in the waterways, the soil at the bottoms and sides of the waterways is loosened by tilling it and then the seeds of grasses/vegetations (preferably seeds of erosion control grasses) are sown. Sometimes, the grass seedlings are directly transplanted or sodding is done at the beds and sides of the waterways. If the moisture content of the soil is not sufficient for germination of the seeds, some light supplemental irrigation may be provided by sprinkler irrigation so that no soil and seeds get washed away. Supplemental irrigation is also required for quick establishment and quality vegetation. It should be kept in mind that the vegetation establishment is perfect before any large flood water enters into the waterways. Special protection such as mulch anchoring is needed during the vegetation establishment period. Where the disturbed soil from the construction areas contributes runoff and sediment directly to the waterways, a 20-ft vegetative filter should be placed along the side of the grassed waterways to filter out the sediment.

8.10.2. Some Important Design Considerations

Following points must be considered before designing grassed waterways.

- Vegetated waterways work better in areas where there are dense stands of sod forming perennial grasses. These grasses permit higher velocity of flow compared to the earthen bare channels without causing any scouring or soil erosion.
- Planting should be established prior to the occurrence of the expected runoff. It is obvious that the grassed outlets get easily damaged by continuous flows over a long period. The waterways may be planted with water tolerant type vegetations. Waterways receiving irrigation tail water will often require mechanical conveyance of sufficient capacity to contain the low-volume sustained flows.
- Waterways need protection from the channel flows until the vegetation is fully established and must have protection from damage by the vehicular traffic and grazing.
- The use of the herbicides that damage the vegetation cover must be avoided.
- Shaping of the waterways must be completed and associated mechanical structures such as drops, pipes etc. must be installed before establishing vegetation.
- The vegetative species chosen must be compatible with the overall crop management system. This is of special importance when selecting perennial type annual species.
- Annual species used for erosion control should be capable of rapid vigorous establishment and growth.
- The grasses in the waterways should be regularly mown with a mower and no grazing should be allowed. Mowing will help to maintain capacity and reduce sediment deposition.

- If trees and shrubs are incorporated, they should be retained or planted in the periphery of grassed waterways. Mid- or tall bunch grasses may also be planted along waterway margins to improve wildlife habitat. Waterways with these wildlife features are more beneficial when connecting other habitat types i.e. wooded tracts and wetlands.
- Provide livestock and vehicular crossings as necessary to prevent damage to the waterway and its vegetation.
- Establish filter strips on each side of the waterway to improve water quality.
- Minimize damage to vegetation by excluding livestock whenever possible, especially during wet periods.
- Inspect grassed waterways regularly, especially following heavy rains. Damaged areas should be filled, compacted, and seeded immediately. Remove sediment deposits to maintain capacity of grassed waterway.

8.10.3. Design of Grassed Waterways

Capacity: The minimum capacity of the waterways shall be that required to convey the peak runoff expected from a storm of 10-year frequency, 24-hour duration. When the waterway slope is less than 1 percent, out-of-bank flow may be permitted if such flow will not cause excessive erosion. The minimum in such cases shall be the capacity required to remove the water before crops are damaged.

Velocity: Design velocity should be calculated using Manning's formula and standard procedures for determining. The Manning's formula is given as:

$$V = \frac{1}{n} R^{\frac{2}{3}} \sqrt{S} \qquad \text{... (8.6)}$$

where, V = velocity of flow in the waterways, m/sec; n = Manning's roughness coefficient; R = hydraulic mean radius defined as the ratio of area to wetted perimeter of flow in the waterways, m and S = bed slope of the waterways, m/m. Values of Manning's n and the limiting velocity in the channel for various bed materials in channel section are given in Table 8.3 (Hansen *et al.*, 1979).

Table 8.3. Values of Manning's n and limiting velocities for different bed materials of channels

Bed materials	Manning's n values	Velocity, m/sec	
		Clear water	Silt mixed water
Fine sand, colloidal	0.020	0.4	0.8
Sandy loam non-colloidal	0.020	0.5	0.8
Silt loam, non-colloidal	0.020	0.6	0.9
Alluvial soil, non-colloidal	0.020	0.6	1.1
Ordinary firm loam	0.020	0.8	1.1
Volcanic ash	0.020	0.8	1.1
Silt clay, very colloidal	0.025	1.1	1.5
Alluvial soil colloidal	0.025	1.1	1.5
Shales and hardpans	0.025	1.8	1.8
Fine gravel	0.020	0.8	1.5
Graded loams to cobbles, non-colloidal	0.030	1.1	1.5
Graded silt to cobbles when colloidal	0.030	1.2	1.7
Coarse gravel, non-colloidal	0.025	1.2	1.8
Cobbles and shingles	0.035	1.5	1.7

The values of the velocity mentioned as above refer to water depth of 1 m/sec. For higher depth, the velocity should be increased by 0.15 m/sec. For canal with sinuous alignment, a reduction of 25% is recommended (Hansen *et al.,* 1979).

The values of R depend on the bed width and depth of flow as well as shape and side slope of the waterways. For example for a trapezoidal channel with side slope Z (side slope, Z = horizontal: vertical) with bed width b and depth of flow d, the value of R is:

$$R = \frac{(b + Z \times d) \times d}{b + 2d \times \sqrt{Z^2 + 1}} \qquad \dots (8.7)$$

Once the values of n, R and S are known, then we can compute the velocity of flow in the waterways. Design velocity with mature vegetation in grassed waterways should not exceed 1.5 m/sec.

Width: Grassed waterways may be parabolic, trapezoidal, rectangular or triangular in shape. The trapezoidal waterways are most common. The bottom width of the trapezoidal waterways should be limited to 15 m unless multiple or divided waterways or other means are provided to control meandering of low flows.

Side Slope: Side slopes should not be steeper than 3:1 (horizontal: vertical) to facilitate in mowing and allow easy movement of farm machineries.

Depth: The minimum depth of a waterway that receives water from terraces, diversions, or other tributary channels shall be that required to keep the design water surface elevation at, or below the design water surface elevation in the tributary channel, at their junction when both are flowing at design depth. Freeboard above the designed depth shall be provided when flow must be contained to prevent damage. Freeboard shall be provided above the designed depth when the vegetation has the maximum expected retardance.

Drainage: Sub-surface drains, underground outlets, stone center waterways or any other suitable measures should be provided in the sites having prolonged flows, high water tables or seepage problems.

Outlets: All grassed waterways should have a stable outlet with adequate capacity to prevent ponding or flooding damage. The outlet may be another grassed waterways or an earthen ditch.

8.10.4. Procedure for Design of Channel Section

The following steps may be used to design stable trapezoidal channel section (Murty, 1998).

Step I : Decide what magnitude of the peak discharge is to be carried by the channel section.

Step II : Assume certain value of permissible limiting velocity to be handled by the channel section and Manning's n (Table 8.3) for particular bed materials

Step III : With assumed velocity, Manning's n and bed/ longitudinal slope, hydraulic mean radius (R) is computed by using Manning's formula (Eq. 8.6).

Step IV : Compute the value of R in terms bed width b and depth of flow d by using (Eq. 8.7) for the given side slope of the channel.

Step V : With the assumed velocity, work out the cross sectional area A of the channel section i.e. A = Q/V

Step VI : Also compute cross sectional area A in terms of bed width b and depth d for the given side slope of the trapezoidal channel section.

Step VII : The two equations in terms of b and d are solved simultaneously to find out the values of bed width b and depth of flow d.

Step VIII : Determine Lacey's f ($f = 2.5\ V^2/R$) to check the stability of the section. The Lacey's f calculated as above should not be less than 1.2.

Sample Calculation: Design a trapezoidal canal section with side slope 1.25:1 to carry discharge of 250 cumecs. Bed slope of the canal section is 1:6666 and Manning's n is 0.018.

Solution: Assume permissible velocity as 1.75 m/sec.

Calculate A by the formula A = Q/V = 250/1.75 = 142.85 m2.

Using Manning's formula and taking n = 0.018, V = 1.75, and S = 1/6666, R computed is 4.15 m.

Calculate P by formula P = A/R
= 142.85 /4.15 = 34.42 m.

Now, $142.85 = b\ d + 1.925\ d^2$ (i)

Also 34.42 = b + 3.85 d (ii)

Solving Eqns. (i) and (ii), d = 6.55 m and b = 9.25 m.

Lacey's $f = 2.5\ V^2/R = 2.5\ (1.75 \times 1.75)/4.15 = 1.84$ which is > 1.2 and hence channel is stable.

So, the bed width of the channel is 9.25 m and depth of flow is 6.55 m.

8.11 Design of Embankment of Farm Pond

The design of the embankment requires the following data:

1. Hydrologic data (stream flow, inflow design flood, sediment, ground water table and water requirement)
2. Climatic data (rainfall and storm intensity, temperature, evaporation rate, maximum and minimum temperature, sunshine hour, relative humidity, wind velocity and direction)

3. Geologic data (aquifers types, presence of impervious / hard sub stratum)
4. Embankment side map (topography, geological features, horizontal and vertical controls, location of existing work at site)
5. Tail water depth (stage-discharge data, stream cross-section and tail and back water curves)

The following parameters are to be considered in design of embankment.

8.11.1. Foundation of Embankment

The foundation where construction work will be made should provide stable support to the embankment and can create a resistance to the passage of water through the dam section. Good foundation materials should provide both stability and imperviousness. A mixture of coarse and fine textured soils like gravel – sand- clay mixtures, gravel-sand-silt mixtures, sand-clay mixtures and sand-silt mixtures will be good foundation materials. Coarse textured materials like gravel and sand provide good support but are highly pervious to water. If such materials exists, precautions should be made to prevent seepage under the embankment. Fine textured materials such as silts and clays are comparatively impervious to water but not very stable materials for the foundation.

The design of cross section depends on the nature of the foundation materials as well as the fill material available at the site. If the material available is of coarse and foundation soil is pervious in nature to water, for earthen embankments in retaining water an impervious core should be provided at the middle of the embankment cross section. If the foundation material is porous in nature, a cut off trench is provided to control seepage. The cutoff trench should be embedded with the impervious soil stratum. The combination of impervious core wall and cutoff trench will be a greater impact on reduction of seepage loss through the embankment and the foundation.

Some times the bottom of the storage area is provided with an impervious blanket where foundation is pervious in nature to control the seepage through the foundation. Fig. 8.2 shows the cross section of core wall with cut off trench and impervious blanket of an embankment.

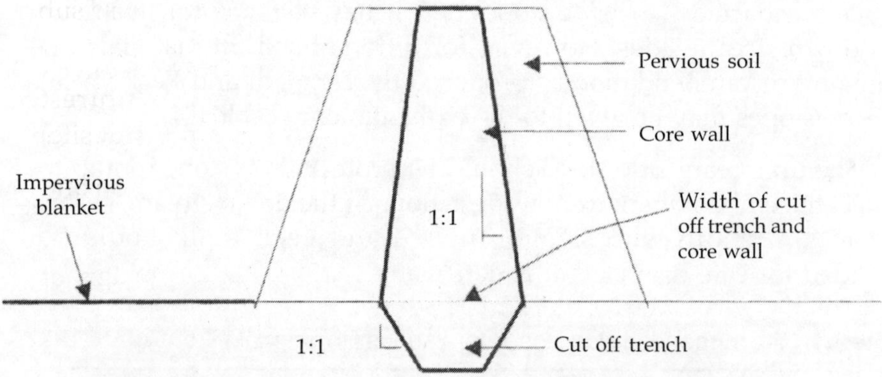

Fig. 2. Impervious core with cut off trench in embankment

The minimum width of core trench is kept 120 cm at the bottom with side slope 1:1 which is provided layer over layer by compaction. The minimum width of cutoff trench at the base should 120 cm with side slope 1:1 (Murty, 1998).

8.11.2. Cross Section of Earthen Embankment

The cross section of earthen embankment depends upon the nature of foundation and fill materials. The evaluation of different components of the embankment section is described below:

Top Width: The top width of the embankment will vary depending upon the height and use of top surface by the public. In general, the top width of the embankment is determined by the following formula:

$$W = 1.105\sqrt{H} + 0.91 \qquad \text{... (8.8)}$$

Where, W = top width of the embankment, m and H = height of the embankment, m.

If the embankment will be used for public passage a minimum top width of 2.5 m may be considered. If it is to be used as a road, a minimum width of 5.0 m or more may be adopted.

Side Slope: The side slopes are dependent upon the height of the dam, nature of the foundation material and fill materials. With average fill materials, the upstream and down stream slope should not be steeper than 3:1 and 2:1 (H:V) respectively. Fill materials of relatively coarse and cannot be compacted well, the upstream and down stream side slopes may be considered as 3:1 to 4:1. Incase of well graded and can be compacted well, side slopes of 2:1 may be taken for upstream and down stream sides. However, for different soil fill materials the following recommendation as proposed by Terzaghi and Peck (1948) for side slopes may be given to the embankment (Table 8.4).

The up-stream side slope should be protected by stone reprap to protect the embankment from wave action. At the down stream, berms of 1 to 3 m width with sloping inward and vegetation should be provided for safe drainage of runoff water.

Table 8.4. Recommended side slopes for various fill materials

Type of fill materials	Upstream slope (H:V)	Down stream slope (H:V)
Homogenous well graded	2.5:1	2:1
Homogenous coarse silt	3:1	2.5:1
Homogenous silt clay or clay	2.5:1	2:1
Sand or sand and gravel with clay core	3:1	2.5:1

Height: The height of earthen embankment is decided after fixing the position of spillway. An additional height and freeboard of 60 to 100 cm above spillway should be given for protection against wave action using Hawksley equation as:

$$h = 0.014\sqrt{D_m} \quad \dots (8.9)$$

where, h = wave height from trough to crest for maximum wind velocity, m and D_m = fetching length or exposure length, m

Determination of net height is regulated by the amount of settlement occurred. The height of settlement depends upon the method of construction and foundation materials used. The following formula may be used for computing the height of settlement in the case of earth embankment (Murty, 1998).

$$s = \frac{e_1 - e_2}{1 + e_1} \times y \quad \dots (8.10)$$

where s = height of settlement, m; e_1 = void ratio before consolidation; e_2 = Void ration after consolidation and y = depth of soil being consolidated

After settlement, generally the free board of 60 to 75 cm is added to the height of the dam.

8.11.3. Seepage through Embankment

Seepage through the earthen embankment should be controlled to ensure the safety of the embankment as well as to minimize the loss of water. When water is standing against the earthen embankment, seepage line or saturation line will exist in the dam section. This is the line below which there will be hydrostatic pressure and above it is negative pressure occurred. On the line, the pressure is atmospheric pressure i.e hydrostatic pressure is zero. It is necessary to predict the position of seepage line in order to (i) ensure that the seepage line should not intersect the down stream face of the embankment which will cause softening and sloughing of the toe (ii) to obtain the dividing line between the wet and dry soil for stability computations and to obtain the top boundary line for draining the flow net for the seepage computations.

The seepage line is affected by the permeability of fill materials and the foundation, the presence of core wall or cutoff wall within the embankment and drainage conditioned in the down stream portion of the embankment . For homogenous fill material, the seepage line cuts the down stream face. The position of the seepage line depends on the geometry of the cross section and not on the permeability of the fill material as long as the section is homogenous. A relatively less pervious materials will take larger times to obtain the final study position of seepage line than a pervious material, but at the end, position of the seepage line in both the cases will be the same provided the flow exists as per Darcy's Law.

Therefore, the discharge per unit width through any vertical section of the embankment is given by (Fig 8.3)

$$q = K\left(h_1^2 - h_2^2\right)/2L \qquad \text{... (8.11)}$$

The above equation does not consider the entrance and exit condition of the line of seepage (h^2 =0). In the absence of tail water, the line of seepage is seen to intersect the impervious base.

Considering an earthen embankment on an impervious base, with no tail water, the length of surface of seepage is given by Schaffernak and Van Iterson (1916) Fig. 8.4 as:

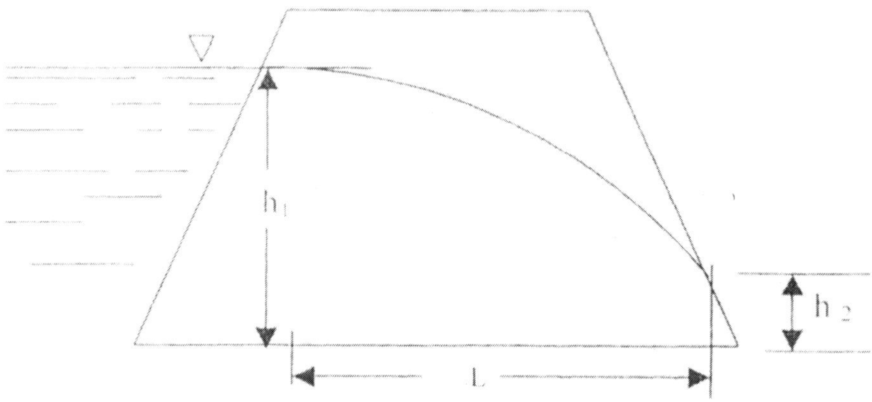

Fig. 8.3. Discharge through embankment

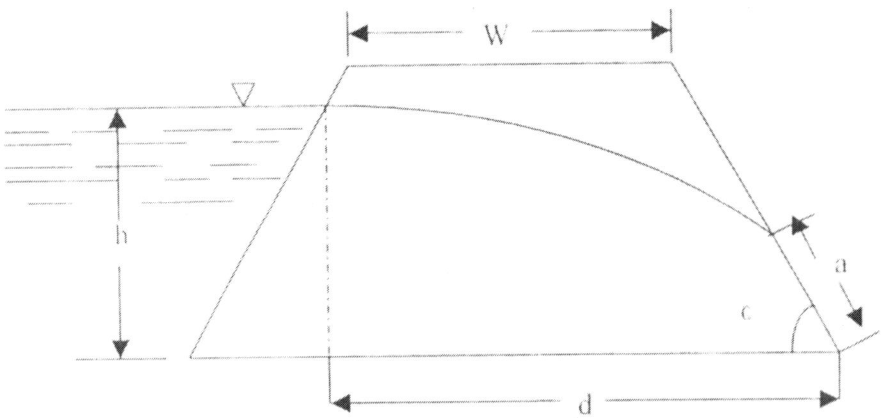

Fig. 8.4. Seepage line determination

$$a = \frac{d}{\cos\alpha} - \sqrt{\left(\frac{d^2}{\cos^2\alpha} - \frac{h^2}{\sin^2\alpha}\right)} \qquad \text{... (8.12)}$$

The above equation shows that the parabolic free surface is tangent to the downstream slope as is required from the stability point of view and from softening and sloughing of the downstream face.

For the entrance conditioned, correction at the upstream side slope is needed. Cassargrande recommended that point D_O instead of point D be taken as the starting point of the seepage line (D_O is 0.3m from point D at upstream reservoir surface). The actual entrance condition is than obtained by sketching DF normal to the upstream side slope and tangent to the parabolic free surface (Fig. 8.5).

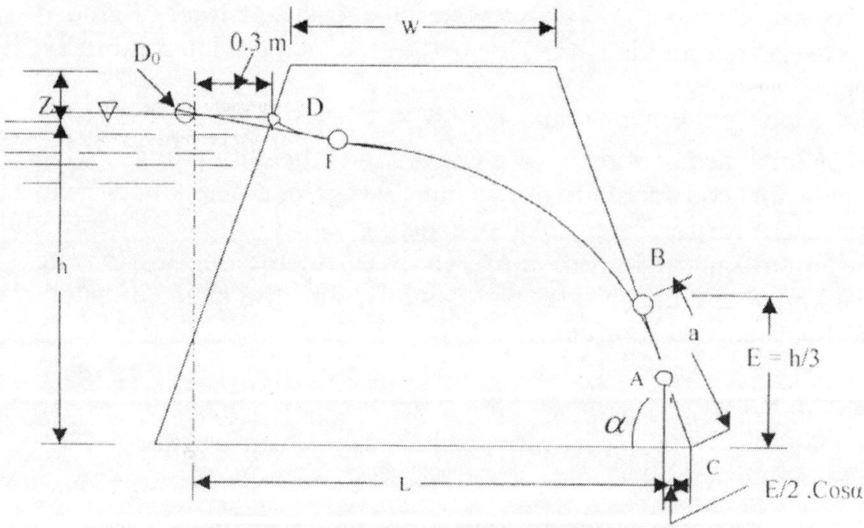

Fig. 8.5. Seepage line position in embankment

In a homogenous embankment, located on impervious foundation where the discharge slopes are faltter than 1:1, it can be assured that the point where the seepage line intersects the down stream face is given (Fig. 8.5) by Eq. (8.13) as:

$$E = a \sin \alpha - h/3 \quad \ldots (8.13)$$

The discharge through a unit width of embankment, q is calculated as

$$q = K (h^2 - e^2) / 2L \quad \ldots (8.14)$$

On substitution of the value of $e = h/3$,q is computed as:

$$q = 4Kh^2 / 9L \quad \ldots (8.15)$$

$$\text{and } L = 1.3\ h + (2z - e/2) \cot \alpha + w \quad \ldots (8.16)$$

where, e = distance from the impervious base to the intersection of the seepage line on the down stream side slope; z = vertical distance from upstream water level to the top of embankment; h = height of water on the up stream side; K = hydraulic conductivity of the embankment material; L= mean length of the seepage line (distance between the starting point of the seepage line D_o and the mid point A of the seepage face BC; w = top width of the embankment and α = the angle of the down stream side and upstream faces.

For safe disposal of seepage water, toe drains or filter of graded and gravels are provided at the toe of the embankment longitudinally.

8.11.4. Stability of Embankment

For small farm ponds having embankment height up to 15 m, the design factors considered in the section "design of embankment" will be sufficient for the safety of the structure. But for large structures with embankment height more than 15 m, the section of the embankment should be designed carefully and the factor of safety should be analysed carefully.

Question Banks

Q1. How should the site clearing be done before construction of farm pond?

Q2. Describe how the demarcation of farm pond area be accomplished.

Q3. What is the use of reference level in pond construction?

Q4. Why is the stepping method of construction in farm pond preferred?

Q5. How should the excavated soils of farm pond be utilized?

Q6. What are the advantages of shoulder bund and toe drain provided in the spoil bank of a farm pond?

Q7. What is a silt trap and how it is useful in the inlet structure of a farm pond?

Q8. Write the formulae to compute the discharge of a broad crested weir with both end contraction and without any end contraction. Explain the various terms used in theses formulae.

Q9. Explain which one of the following two types of weirs in the inlet section of a farm pond will measure higher discharge under same length of crest of the weirs and same height of water flowing over the crests.
(a) Rectangular weir and (b) Trapezoidal weir

Q10. A rectangular weir is used as inlet to a farm pond and it has crest length 50 cm and the height of water flowing over the crest of the weir is 30 cm. Determine the discharge passing over the weir when (i) there is no end contraction (ii) when there is only one side end contraction and (iii) when both sides of the weir have end contraction.

Q11. A trapezoidal weir has top width 100 cm and bottom width 40 cm. It has side slope 1:1. If the water is flowing full in the weir, then determine the discharge flowing through the weir.

Q12. Write the different steps required for the installation of a weir in the inlet section of a farm pond.

Q13. What is a grassed waterway and what are the advantages of digging a grassed waterway in farm pond areas?

Q14. Describe the various design procedures of establishing a grassed way in a farm pond area.

Q15. Design a trapezoidal canal section with side slope 1.2:1 to carry discharge of 200 cumecs. Bed slope of the canal section is 1:6000 and Manning's n is 0.016. Assume the permissible velocity in the channel as 1.8 m/sec.

Q16. How is the vegetation in a grass waterway established?

Q17. What are the uses of embankment of a farm pond? What data are required to be collected before designing the embankment of a farm pond?

Q18. Describe how you can find out the design dimensions of embankment of a farm pond.

Q19. Draw a schematic presentation of farm pond embankment and show the seepage line through it.

Q20. How can the seepage through a farm pond embankment be checked?

Q21. Write about good foundation requirements of a farm pond embankment.

Q22. Describe the procedures to check the stability of an embankment of a farm pond.

References

Belgaumi, M.I., Itnal, C.J. and Radder, C.D. 1997. Farm Ponds. Technical series 4, Publication Centrte, University of Agricultural Sciences, Dharwad, pp. 47.

Coche, A.G., Muir, J.F. and Laughlin, T. 1995. Simple Methods for Aquaculture Pond Construction for Freshwater Fish Culture Building Ponds, FAO of the United Nations, Rome, pp. 355.

Hansen, V.E., Israelsen, O.W. and Stringham, G.E. 1979. Irrigation Principles and Practices. John Wiley and Sons Inc., New York.

Lenka, D. 2001. Irrigation and Drainage. Kalyani Publishers, New Delhi.

Murty, V.V.N. 1998. Land and Water Management Engineering. 2nd Edition, Kalyani Publishers, New Delhi, pp. 586.

Senapati, S.C. 2008. Practical Manual on Soil and Water Conservation Engineering. College of Agricultural Engineering and Technology, Orissa University of Agriculture and Technology, Bhubaneswar, India.

Sharma, R.K. and Sharma, T.K. 2002. Irrigation Engineering including Hydrology, S. Chand and Company Ltd. New Delhi.

Terzaghi, K and Peck, R.B. 1948. Soil Mechanics in Engineering Practice. John Wiley and Sons. New York.

Chapter 9

Losses of Water in Farm Pond

9.1 Introduction

The water harvested in the farm pond is a very valuable commodity. Every care should be taken so that not a single drop is wasted. The two major forms of losses in farm pond are (i) evaporation losses and (ii) seepage losses. Evaporative water loss (E_{loss}) and losses due to seepage are the two vital natural outflow components of unlined farm ponds. Studies conducted by Guerra *et al.* (1990) and Pal et al. (1994) indicated that the seepage loss from unlined ponds amounts to 45-67 per cent whereas the E_{loss} constitutes 30 per cent of the total out flow from the system. Water losses by evaporation and seepage were 5.3 and 9 mm day^{-1}, respectively amounting to 70% of total storage in the ponds (Syamsiah et al. 1994).

Estimate of evaporation across the country indicates that it is as high as 2000 mm in the semi-arid tropics. Annual average value of the evaporation loss varies from 1400 to 1800 mm across major parts of the country: the value being highest in west Rajasthan, parts of Saurasthtra and Tamilnadu and lowest in coastal Mysore, Bihar plateau and east M.P. Reliable statistics reveal that about 70 M ha m of water evaporates from the water bodies and land surfaces out of the total annual precipitation of about 400 M ha m received by the country. The total evaporation losses from the water surfaces are about 5 M ha m from the total storage of 15 M ha m in the reservoirs including ponds, lakes spread all over the country (Belgaumi et al., 1997). In this section, the methods to estimate the losses like seepage and evaporation and various techniques to control them are discussed.

9.2 Evaporation Losses in Farm Pond

Evaporation is the process of conversion of water from liquid state to gaseous state and its transfer into atmosphere in this form. It is a natural process and occurs when solar radiation falls on a water body and energy so transmitted is acquired by water molecules which in turn gets separated and move upward causing evaporation. In the hydrologic cycle, evaporation plays a vital role in causing precipitation. Thus, in one hand it is required to make precipitation which builds up the water resources and on the other hand it causes huge losses of the precious harvested water of the farm pond and other reservoirs. The high costs involved in the construction of storage structures including farm ponds, make it imperative to utilize every single drop of stored water. Evaporation loss from the small and shallow ponds is higher than the large and deep ponds and may be as high as 50% of the capacity of the pond. Evaporation loss in ponds is higher in summer months from March to June and lower in winter months from October to February.

Evaporation losses depends on the water spread area of the pond and the rate of evaporation which depends on the climatic factor like temperature, wind velocity, relative humidity of air, atmospheric pressure as well as nature and size of the water body. Water spread area of a particular size and shape of pond with a given side slope varies with depth of water in the pond.and hence the evaporation loss of the pond varies daily. The computation of water spread area otherwise called as the top water surface area for an embankment type farm pond is given by the Eqns. (6.14 and 6.15) in Chapter 6. A sample calculation of water spread area of embankment type pond for two sizes i.e. 10 and 12% of the farm area of 800 m^2 with 1:1 and 1.5:1 pond side slope is presented in Table 6.2. The nomograph (Fig. 6.6) helps in finding out the water spread area on any day having a particular depth of water in the pond on that day. It is important to note that the nomograph presented in this book as Fig. 6.6 is only meant for 800 m^2 farm area and for a square shape embankment type pond having side slope 1:1 and sizes varying from 6 to 20% of the farm area only. The procedure of determining the water spread area on any day is described as follows.

Suppose on any day, the depth of water in the pond is 1.5 m. The pond used is 12% of the farm area and the total farm area is 800 m^2. Depth – volume relationship of the pond at 1.5 m depth gives pond storage (volume) as 40 m^3. From the volume-water spread area/top water surface area at 40 m^3 pond storage we get pond top water surface area 41.1 m^2 (Fig. 6.6). Multiplying the evaporation loss of pond on that day with the so calculated water spread area, we can determine the volume of evaporation loss of the pond.

The various methods of determining rate of evaporation losses are:

9.2.1 Analytical Method

There are two analytical methods to compute evaporation losses. They are water budget method and mass transfer method.

9.2.1.1 Water Budget Method

This is based on water balance approach in which various inflows and outflows to and from the farm pond are considered. The equation used to evaluate evaporation is:

$$E = S_1 + R + SR - Is \pm U - S_2 \qquad \text{... (9.1)}$$

where, E is evaporation, S_1 and S_2 are initial and final storage, R is rain fall, SR is runoff / surface inflow to the pond, Is is the supplemental irrigation/surface outflow from the pond and U is the underground inflow or outflow into or from the pond. The various terms used in above equation are expressed in terms of depth unit, generally in cm.

Surface run off generated from the crop field at the upstream and direct rainfall are two inflow components of the pond. The storage depth in it increases during rainfall events and decreases at irrigation events. A graduated staff gauge can be installed at the center of the pond to take care of the full depth of storage up to embankment height. The depth of storage is used to compute the daily water spread area in the pond. The seepage loss from the completely plastic lined farm pond may be assumed zero. So, the change in the depth of storage during an interval suggests the volume of water lost due to evaporation from the farm pond only.

9.2.1.2. Mass Transfer Method

The evaporation in this method is computed as:

$$E = \frac{-10K^2\rho(q_2 - q_1)(u_2 - u_{1)}}{\left(\ln\left\{\frac{Z_2}{Z_1}\right\}\right)^2} \quad \text{... (9.2)}$$

where, E = evaporation, mm; K = von Karman constant ≈ 0.41; λ = density of air, (g/cc); q_2 and q_1 are specific humidity at heights Z_2 and Z_1, respectively that are taken as 2 m and 1 m, respectively above the water surface; and U_2 and U_1 are wind speeds (cm/s) at heights Z_2 and Z_1, respectively.

9.2.2 Empirical Method

There are several empirical methods available for computing evaporation loss. Most of these methods take into account the evaporation loss of the farm pond which in turn depends on temperature of air water interface, wind velocity, relative humidity and saturated vapour pressure of air.

9.3.3 Energy Budget Method

In this method, evaporation is given as:

$$E = \frac{R - G}{L}\left(\frac{1}{1+\beta}\right) \quad \text{...(9.3)}$$

where, $R = H + LE + G$ and β is given as:

$$\beta = \frac{T_s - T}{e_s - e} \times 0.61 \times 0.001 P$$

In the above equations, E = evaporation (mm); R = net radiation flux received at the surface (cal/cm²); H = sensible heat flux (cal/cm²); LE = latent heat flux, where L = latent heat of vapourisation (cal/g); E as defined above; G = heat flux in the soil or water (cal/cm²); β = Bowen ratio; Ts = temperature of water surface (°C); T

= air temperature (°C); Is = saturation vapour pressure at Ts in hectopascal (hpa); e = vapour pressure of air at T in hpa; and P = station level pressure in hpa.

Based on Penman's equation (1948), the Kohler-Nordenson-Fox equation (Pochop et al., 1985) describes evaporation from water bodies as the combination of water loss due to radiation heat energy and the aerodynamic removal of water vapor from a saturated surface.

The general form for the combination equation is:

$$E = \frac{\Delta}{\Delta + \gamma} Rn + \frac{\gamma}{\gamma + \Delta} Ea \qquad \text{... (9.4)}$$

where, E = evaporation in inches per day; Δ = slope of the saturation vapor pressure curve at air temperature in inches of mercury per degree F; γ = psychrometric constant in inches of mercury per degree F; Rn = net radiation exchange expressed in equivalent inches of water evaporated; and Ea = an empirically derived bulk transfer term. Ea is expressed as:

$$Ea = f(u)(e_s - e_d) \qquad \text{... (9.5)}$$

where, f (u) = wind function; and (es - ed) = vapour pressure deficit.

So, the evaporative water loss from water body can be checked by obstructing the radiation heat energy coming on to the water surface and further reduction can be achieved by lessening the vapour pressure deficit or the wind speed in the microclimate.

9.2.4 Pan Evaporation Method

In this method, the evaporation loss measured by an evaporimeter located near the pond is measured. The ratio of the evaporation loss measured by the evaporimeter to the pond evaporation loss on annual basis is a constant and hardly varies from region to region. However, on short term period basis, the ratio depends on a number of factors like heat storage of the evaporimeter/pan, size of the pan, water depth in the pan, colour of pan etc. The evaporation loss from a pan is greater than that of the pond. As the size of the pan increases, the difference of the evaporation loss of both pan and pond decreases. The rate of evaporation, in general, is observed to be more for deep buried pan with higher depth of water than shallow pans with less depth of water.

Similarly, the evaporation from dark pans is greater than that from unpainted, galvanized pans. The white coloured pan shows the lowest evaporation. Covering the pan with a screen helps to reduce the pan evaporation to the equivalent of that from a reservoir. The screen helps to maintain the pan temperature uniform throughout day and night and reduces turbulence at the water surface so as to maintain uniform rate of evaporation.

9.2.4.1 Selection of Site of Evaporimeter

While selecting the site of evaporimeter near the farm pond, care should be taken so that:

- It is very near to the pond.
- The evaporimeter must be placed on a level ground.
- There should not have any obstruction for evaporation of water from the pan. If there are any tree or building near by, then the pan should be placed at the location which is at least 10 times more then the height of the tree or building.
- The pan should not get submerged due to heavy rain or flood.
- It should be located upwind of the pond.
- The highest ground water level must be more than 2 m from the ground surface.
- Moreover the site of evaporimeter must be easily accessible.

9.2.4.2 Types of Evaporimeters

Evaporation pan is an instrument for measuring the amount of water lost by evaporation per unit area in a given time interval from a shallow container. The commonly used evaporation pans are given as follows (Sharma and Sharma, 2002):

(i) B.I.S. Class A Pan (Modified)

Class A pan standardized by Bureau of Indian Standards (B.I.S.) and used in India consists of a large cylindrical pan of 1220 mm diameter and 255 mm deep. It has a fixed point gauge. It is made up of 1 mm thick copper sheet tinned inside and painted white outside. It is covered with a wire mesh. The pan rests on a white painted wooden grill of 100 mm above the ground. A stilling well is provided near the fixed point gauge for providing undisturbed water surface near the point gauge. A thermometer is fixed to the side of the pan to measure surface temperature of water. Fixed point gauge is used as a reference point for free water surface in the pan. A graduated cylinder is used to measure the volume of water to be added to the pan that is lost by evaporation in a given time so that the water level reaches the reference point. Dividing the volume of water so added with the pan surface area gives depth of evaporation loss from the pan. If there is any rainfall during the time period in which evaporation loss is measured, then due allowance is made from the catch of nearby raingauge. Dividing the depth of evaporation loss with the time period during which evaporation measurement is taken, evaporation rate can be found out. The coefficient used to obtain evaporation loss from free water surface area of the pond from pan evaporation may vary between 1.10 - 0.90 for pond evaporation of the order of 4-5 mm/day, between 0.75 - 0.65 for pond evaporation over 10 mm/day and 0.80 for evaporation between 5 -10 mm/day.

(ii) Class A U.S. Weather Bureau Land Pan

It is a circular pan made of galvanized iron or monel metal. It is 1210 mm in diameter and 255 mm in depth. The pan rests over a wooden open platform 150 mm above the ground. The platform allows free wind circulation below the pan. The pan is filled with water within 50 mm from the rim of the pan. A hook gauge is used to measure the evaporation loss of water from the pan. The coefficient of 0.7 is used to convert the pan evaporation to the pond evaporation.

(iii) Colorado Sunken Pan

It is a unpainted galvanized iron of 914 mm square and 457 mm deep. The pan is sunk in the ground till the rim remains 51 mm above the ground surface. The water level is maintained at or slightly below the ground level in the pan. A hook gauge is used to measure the depth of water lost as evaporation from the pan. The measurement can also be made with the help of fixed point gauge. The hook can read up to 0.02 mm. The coefficient used to convert the pan evaporation to pond evaporation is 0.89.

(iv) US Geological Survey Floating Pan

It is developed to simulate the conditions similar to those of the surrounding water in the reservoir. The pan is made of galvanized iron. It is 900 mm in square and 450 mm deep. It is supported on drum floats in the centre of a raft 4.25 x 4.87 m and kept afloat in the pond. The water level in the pan and that in the surrounding is kept at same level. This pan is not generally used. If it is used then, Class A U.S. Weather Bureau Land Pan is also installed near the floating pan to supplement the data during the missing period. The floating pan is less accessible for taking readings. Sometimes wind affects the accuracy in measurement since the waves formed by wind may cause splashing of water. The coefficient used to convert the pan evaporation to pond evaporation is 0.8.

(v) Class A U.S. Weather Bureau Floating Pan

It is a land pan but is not rested over wooden open support. The water level inside the pan is 75 mm below the top of the pan. The water level both inside and outside are kept at same level. A fixed index point in the centre of the pan is marked. The quantity of water added to the pan to bring the water level reading up to this fixed index point gives the amount of water lost as evaporation from the pan.

It is observed that the evaporation loss in hot months of May and June is 2-5 times more than that in winter months of December and January. For the Indian subcontinent, annual evaporation loss from the reservoir ranges from 150 – 200 cm. A maximum level of 300 cm can evaporate in Rajasthan and a minimum of 50 cm can evaporate in Jammu and Kashmir.

Sample Calculation: An evaporation pan is 1.5 m in diameter and is used to measure evaporation loss of a farm pond. The initial and final water level reading in the pan are found to be 10 cm and 11 cm, respectively. A rainfall of 4 cm has occurred during the period the observation was taken. To keep the water level same in the pan, 2 cm depth of water has been removed from the pan. Find out the pan evaporation. If the pan coefficient used to convert the pan evaporation to pond evaporation is 0.8, find out the pond evaporation also.

Solution: Initial water level in the pan = 10 cm.

Rainfall during the observation period = 4 cm.

Water removed from the pan = 2 cm.

Net addition of water in the pan = 4 – 2 = 2 cm.

Had there been no loss, the water level in the pan would have been = 10 + 2 = 12 cm.

Final water level in pan = 11 cm.

Hence, evaporation loss of the pan = 12 – 11 = 1 cm.

Pan coefficient given = 0.8

Hence, pond evaporation loss = 0.8 x 1 = 0.8 cm.

9.3 Reduction of Evaporation Losses

Storage of irrigation water in a pond is a very valuable commodity. Appreciable evaporation loss from large water spread area of a pond has a consequential loss on power generation, irrigation supply and many other associated activities. Some of the methods used for reducing evaporation losses are:

9.3.1 Reduction of Pond Surface Area

Evaporation is a surface phenomenon and hence if the surface area of the pond is reduced, then their evaporation loss is considerably reduced. A pond with greater depth of stored water having smaller surface area has less evaporation loss compared to a shallow pond with large surface area; both the ponds having same volume of stored water.

9.3.2 Covering Pond Open Surface with Biological Shades

Biological shades like bottle gourd (*Lagenaria siceraria*) pumpkin, bitter gourd, cucumber etc can be laid over the open water surface of a farm pond with the support of a wooden platform made up of bamboo sticks and other materials. The creepers of the gourd when moves with tendrils spreading both in forward and lateral directions, effectively creates a shading barrier and bares the direct sun light to fall on pond water. This process reduces the evaporation of pond water and thereby creates a large storage in the pond. Cooley and Idso (1980) observed 2.9 per cent reduction in E_{loss} when lily pads were used to cover the water surface. Wardana et al. (1996) found that shading of water surface by cover crops (squash) in combination with tree crops (banana) for wind breaking could reduce evaporation rate by 50%. However, the disadvantages of growing the floating aquatic plants in farm ponds are that they themselves consume a lot of water from the pond to meet their transpiration requirements and thus have adverse affect on storage of water in the pond.

9.3.3 Monomolecular Film

Chemicals either in the form of powder or solutions may be spread over the water body which reflect away energy input of the sun as a result of which evaporation is reduced. Alcohol such as cetyl alcohol, also called as hexadecanol and stearyl alcohol (octodecanol) are known to form a monomolecular film on contact with water which is sufficiently enduring during field conditions. The invisible film is non-toxic and reduces evaporation by about 50-60% at an average wind speed of 16.55 km/hr and at the same time helps as a free passage for rain, oxygen and sun light (Cooley, 1974). However, the monomolecular films have the disadvantages that they get diluted in water quickly and then become ineffective to reduce the evaporation of water.

9.3.4 Wind Breaks

It is a well known fact that the turbulence affect of the water surface caused due to wind action increases the evaporation loss. Hence, wind breaks like growing tall trees around farm pond can minimize the turbulence affect and hence has an indirect affect on reducing the evaporation loss to some extent. But, these wind breaks are useful only for small ponds. It is found out that a reduction of wind velocity by 25% reduces the pond evaporation loss by only 5% and thus is not significant on sustained basis.

9.3.5 Locating Ponds at High Altitudes

When the ponds are located at high altitudes, then the evaporation is considerably reduced. This is due to the reduction in dew point temperature and reduction in surface water temperature at higher altitudes.

9.3.6 Covering Pond Water with Shading Materials

A number of materials can be used as effective barrier to reduce the impact of direct sun light to fall over the pond water surface and hence reduce the evaporation loss. Shading materials like plastic films, thatches, paddy straw, sugarcane trash etc. reduces the radiation affects on water surface and decreases the evaporation loss. The amount of reduction of evaporation from the pond depend on the nature of the shading materials and the percentage of coverage of the materials on the water surface. If the coverage is 100 percent, then the amount of evaporation reduction will be maximum and thus, partial coverage will have reduced percentage of evaporation. Some other shading materials are dye mixed with pond water, coloured pan, plastic mesh and sheet, polystyrene beads and sheet, white spheres, white butyl sheet etc. Out of these measures, polystyrene raft, plastic sheet and foamed butyl rubber are the best measures since they are reported to reduce the evaporation loss by more than 90% (Fig. 9.1). Field experiments conducted at IIT, Kharagpur reveal that the evaporation loss from an open farm pond constitutes 14.5% of the total outflow whereas a pond completely covered with plastic films on the top reduces the evaporation loss to 3.4% of the total outflow of the pond (Panigrahi, 2001). It is found from experimental studies that paraffin oil containing spreaders of high molecular weight may reduce the evaporation loss up to 15%.

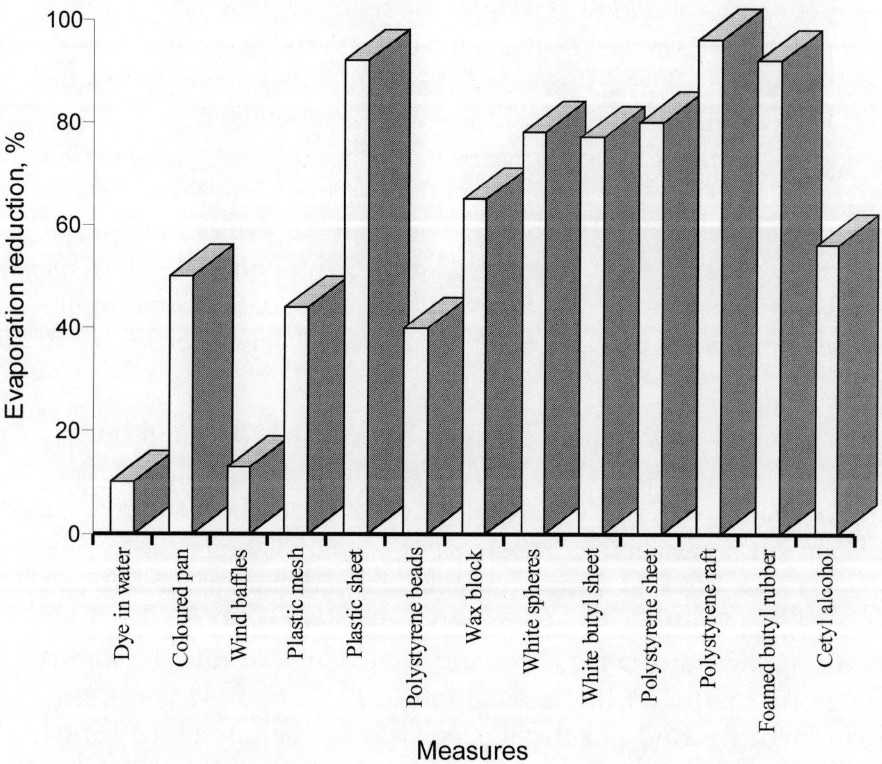

Fig. 9.1. Effectiveness of various evaporation reduction measures

9.4 Seepage Losses in Farm Pond

Water stored in a farm pond is a costly commodity and care should be taken to store it for the maximum period as possible with minimum loss. Most of the farm ponds used for irrigation are unlined which cause huge losses due to seepage. This loss is significant when the pond is underlain by porous strata or when the bed material of the pond consists of coarse textured soil. Small farm ponds constructed in coarse textured soils; especially in arid and semi-arid regions are found to get dry completely just after the rainy season is over. However, ponds constructed in heavy soils are found to have less seepage losses. In general the seepage loss in unlined small farm ponds range from 40-60% depending on the water table position below the ground surface, types of soils in which the pond is constructed and hydraulic gradient available between the pond water level and water level of adjacent areas. It is observed that the seepage loss in newly constructed pond is very high and as time progresses, siltation of the pond reduces the seepage loss. Seepage loss in a pond depends on the wetted surface area of the pond which in turn depends on the depth of water in the pond. The wetted surface area of the embankment type farm pond on any day is estimated by the Eqns. (6.16 and 6.17) as discussed earlier in section "Dimensions of Farm Pond" in Chapter 6. Procedure to estimate wetted area of the pond for two sample size farm pond i.e. 10 and 12% sizes for farm area of 800 m^2 with 1.5:1 and 1:1 side slope has been discussed earlier and computed values for different depths have been presented in Table 6.3 in the section "Dimensions of Farm Pond" in Chapter 6.

The wetted surface area of an embankment type pond on any particular day having a given depth of water can be determined by using nomograph (Fig. 6.6). Suppose on any day in a square size embankment pond having 12% size in a farm of 800 m^2 with 1:1 side slope is 1.5 m. We find out the volume of pond storage for 12% size pond at 1.5 m depth as 40 m^3 and pond water spread area as 41.1 m^2. Now using the water spread area and wetted area relationship of the nomograph (part C of the nomograph of Fig. 6.6) at 41.1 m^2 water spread area, we find out the wetted area of the pond as 59.5 m^2. Multiplying the seepage loss in unit of depth of water with the wetted area of that particular day, we can get the volume of seepage loss of stored water of the pond.

Seepage loss in a farm pond can be computed by water balance model Eq. (7.20) in which the inflows including the rainfall and runoff and outflows including the evaporation loss, water diverted from the pond for supplemental irrigation and other uses are measured.

9.5 Methods to Reduce Seepage Losses

Since, irrigation water is very costly, efforts should be made to reduce the seepage loss of the pond. Seepage loss can be reduced broadly by two mechanisms. They are (i) reducing wetted surface area of the pond and (ii) using a cost effective sealant.

9.5.1 Reducing Wetted Surface Area of Pond

Seepage increases with the wetted area of a pond and so special considerations during design must be given to minimize the wetted area per unit of storage capacity. This may be achieved by increasing the side slope i.e. steepening the sides of the pond as well as increasing the depth of the pond. Cluff (1977) has suggested compartmentalized pond in which the pond is divided into two or more sections and as one section is emptied, the water is pumped out from another section. By doing so, the wetted area and top water surface area of the pond can be considerably reduced and so seepage and evaporation loss of water in the pond can be reduced to a great extent.

9.5.2 Use of Sealants

A newly constructed pond has self sealing property by silting. The runoff of the catchment carries some silt and clay which gets deposited in the side and bottom of the soil in the pond and reduces the pore spaces of the soil. This, consequently results in the reduction of the seepage loss of the pond. Studies at Dehradun and Rajkot (Vijayalakshim *et. al.,* 1982) reveal that seepage from a newly dug out pond get stabilized to a very low rate in a period of 8 years due to silting. Silting also reduces seepage rate in brick lined pond (Verma *et. al.,* 1984). In some cases, puddling caused due to trampling by animals causes the compaction of pore spaces of the soil in the pond which has a consequential affect in checking seepage losses in the pond.

A number of sealants/ lining materials are now available to reduce the seepage loss. Lining of the pond though costly can reduce the seepage loss and improve the effective storage of the pond. Different lining materials used to reduce the seepage losses are plastic film, soil cement lining, bitumen lining, clay lining, cow dung lining, brick cement lining etc. Lining for reduction of seepage losses is feasible only for small pond. The permanent and most effective and costly is brick masonry or cement mortar lining. Descriptions of a few lining materials used for seepage control in the pond are presented below.

9.5.2.1 Clay Lining

Natural clay can be used for lining with varying degree of efficiency, especially when lower cost is desirable. Clay lining can be applied by two methods; (i) by placing a blanket of relatively impervious clay 15-30 cm thick over or within the permeable bed and sides of the pond and (ii) by dispersing clay in the water and having it filtered out and seal off the pores in the permeable lands. Shally (1965) observed that in some parts of Bengal and Madras, the seepage losses by heavy clay soil are so effective that lining is hardly necessary or justified. Water losses through clay is much less than bricks and is as or more pervious than ordinary cement concrete. Kamra and Dhruvanarayana (1987) reported that alkali soils having poor infiltration rates can be preferred for seepage control in farm ponds if available in the vicinity of the pond. Biswas and Mallick (1997) observed that the burnt clay tiles used as a lining material can reduce the seepage losses and the saving of water due to seepage by it is about 98.8% more than the earthen materials. The average seepage rate by the burnt clay tiles was observed to be 1.54 $cm^3/cm^2/day$. Hegade et al. (1981) reported that the seepage losses using soil cement plaster (5:1) is ideal from the combined considerations of the cost and efficiency. This lining is made up of mixture of cement and natural soil. For good results the mixture should be mixed well, laid and compacted. It should be cured for seven days with moist soil cover. The drawback of the lining material is that it is not weather resistant and its life is comparatively short and the repair and maintenance costs are relatively higher. It also does not permit higher velocity of flow even that in an unlined channel.

9.5.2.2 Cement Concrete Lining

Structures lined with cement concrete are stronger and more stable than lined with any other material. It gives long service. Though the initial investment for lining is more, later on the repair and maintenance costs are less. Concrete mixture usually recommended for lining is 1:3:4 (cement:sand:gravel) and thickness for lining is 4-5 cm (Biswas, 2008). The sides and bottom of the structure should be compacted at suitable moisture content. When concrete hardens, it shrinks and cracking may result. Joint must be provided at a distance not more than 2 m in order to localize and control the cracks. Cement concrete lining can withstand higher velocity more than 2.5 m/sec because of its greater resistance to erosion and is therefore preferable than any other type of lining. Murthy (1988) have reported the effectiveness of different lining materials for seepage control which is presented in Table 9.1.

9.5.2.3. Asphalt Lining

Asphalt is a petroleum product which when mixed with sand and gravel is used as a lining material in the same way as cement. Asphalt concrete lining is cheaper than cement concrete liming. According to Chowla (1992), the life of asphalt lining varies from 10-20 years. Locally available bituminous and other low cost indigenous materials were used for lining under field conditions in Bangladesh by Khair et al. (1984). The following two types of prefabricated asphalt melts have shown promising results in seepage control in Bangladesh. They are (i) Gunny (coarse sack cloth made of jute) reinforced asphalt melt and (ii) Synthetic cloth reinforced asphalt melt. The results have indicated that the former lining material reduced the seepage losses considerably.

Table 9.1. Effectiveness of different lining materials for seepage control

S. No.	Lining materials	Seepage loss, lit/hr/m²
1	Control (No lining)	18.56
2	Cowdung + Paddy husk + Soil plaster (1:1:10)	16.98
3	Cement plaster at bottom (1:6)	12.99
4	Cement + Soil plaster (2:10)	0.85
5	Polythene sheet	0.32
6	Paddy husk ash plaster	11.60
7	Coastal saline soil plaster	5.47
8	Fly ash + Sand plaster (1:1)	2.5
9	Clay	12.07

9.5.2.4. Brick Lining

Brick lining involves the joining of the bricks with soil cement plaster and is relatively requires less investment compared to cement concrete ling. It is easier to handle and requires less technical knowledge. These types of lining involve low consumption of cement and can be laid without use of any machinery. However, it is not that effective in seepage control as compared to cement concrete lining.

9.5.2.5. Bentonite Lining

Seepage losses in a well graded coarse soil may be reduced by mixing bentonite in the surface soil of the pond. Actually bentonite can absorbe several times more water and swell 8-15 times its volume. The main advantage of bentonite lining is low cost and simplicity of installation while its disadvantage is its short life. Bentonite loses its effectiveness and durability due to fluctuation resulting from wetting and drying of the pond and removal of surface layer by erosion and seepage. Detailed studies on bentonite for seepage in ponds are carried out in western U.S. About 5 kg of bentonite per square meter of soil is required (Bowman, 1967; Shen, 1959; Rollins et al, 1970).

9.5.2.6. Alkali Soil Lining

Preliminary investigations in small ponds reveal that it is practicable to reduce seepage loss effectively by plastering the pond with alkali soil of 5 cm thickness (Venkatanadhachari et al., 1979). Alkali soil has also been used for pond lining at Ahantpur (Vijayalakshmi et al., 1982).

9.5.2.7. Soil Deflocculents

Soil deflocculents like sodium carbonate, sodium chloride, tetrasodium polyphosphate and sodium tripolyphosphate can be used for reducing permeability of pond surfaces. Sodium chloride and sodium polyphosphate failed within 6 and 8 months, respectively, but sodium carbonate was more durable. It was suggested that for successful sodium carbonate treatment at least 3 cm of soil should be put over the pervious materials and the clay content of soil should be at least 15% and cation exchange capacity should be greater than 15 mil-equivalents per gram (Reginato et al., 1973).

9.5.2.8. Gleization

Gleization is a new method used for reducing seepage in ponds. It involves the use of organic matter like straw and manure. Under the anaerobic condition, the metabolic activities of certain anaerobes affect a chemical reduction process in soil which results in increased plasticity and dispersity and a decrease in permeability (Nicholaichuk, 1978).

9.5.2.9. Chemical Sealants

U.S. Bureau of Reclamation (1963) studied many chemicals including resins, silicones, linings but none was found suitable in seepage control. Even cationic asphalt emulsion, petroleum emulsion and resinous polymers were tested and found to be short livid and affected by wetting and drying and erosion.

9.5.2.10. Enzymes

Some enzymes have also been tested for seepage control in ponds. Bayer and Cluff (1971) mixed organic enzyme with lubricants as aids in compaction of pond surface and found reduction in seepage rate.

9.5.2.11. Polythene Lining

Polythne films have been widely used for lining purpose in irrigation channels and farm ponds by many researchers (Panigrahi, 2001). These lining can reduce the seepage losses to the minimum even zero if perfectly laid. The percentage of seepage loss by it is maximum than any other lining material. Though the initial investment for lining the structure is more, yet compared to the enormous amount of saving of costly irrigation water, the benefit cost ratio is found to be maximum. While lining the pond with polythene, cares should be taken so that there is no puncture or any gap between the cut pieces of the sheets. The sides of the pond should be constructed with steps (Fig. 9.2) and over the steps, the polythene sheet is laid. Then the soil is covered so that the sheet remains buried in soil. Then it is compacted and the sides of the pond is made sloppy as in the original pond. By doing so, the wears and tears of the polythne sheet is reduced and its longesvity increase. If properly laid, the life span of a ploythene lined pond can go for 10-15 years. Fig. 9.3. gives the overall view of a plastic lined pond full with harvested water.

Fig. 9.2. Steps constructed in sides of ponds for plastic lining

Fig. 9.3. A view of plastic lined pond full with harvested water

Various lining materials have been tried at different centres of All India Coordinated Research Project for Dry Land Agriculture for seepage control in ponds. Results of the effectiveness of the materials are presented in Table 9.2 (Vijayalakshmi et al., 1982). It is observed that of the different types of linings plastic overlaid by brick as well as brick lining overlaid by cement plastering have tremendous affect on reducing the seepage losses in the farm ponds. They reduce 100 percent seepage loss in the pond and thus are very effective in lining the ponds.

However, it is important to note that before using any lining materials for seepage control in the farm pond, the economic analysis relating to life of the pond and cost of lining materials of the pond must be taken into account. At the same time, the amount of saving of irrigation water by reduction of seepage and its use in increasing the crop productivity, total crop production and other associated benefits must also be considered. Finally with present worth analysis the different economic indicators like net profit, benefit-cost ration, internal rate of returns and pay back periods are to be worked out. The lining material that gives more net profit, benefit-cost ratio, and internal rate of returns and less pay back period is the best from seepage control point of view of the pond. Detail procedures of economic analysis of farm pond irrigation systems based on present worth analysis are presented in Chapter 11.

Table 9.2. Results of effectiveness of some sealants for seepage control

Research centre	Material used	Seepage as percentage of control
Bangalore	Clay + Na Cl + Na_2CO_3 (20:5:1)	19
	Soil cement (5:1)	30
	Soil cement (10:1)	42
Dantiwada	Plastic overlaid by brick work	9
	Lime mortar (1:6) with asphalit lining	11
	Cement sand (1:6)	19
Hyderabad	Plastic overlaid by brick work	0
	Brick lining overlaid by cement plastering	0
	Asphalt	13
Ludhiana	Bottom lined by polyethylene	2
	Sides lined by brick lining	6
Rajkot	Soil + cow dung + straw (7:2:1)	11
	Soil compaction to high bulk density	43
Ranchi	Coaltar	44
	Clay	56
Varanasi	Black polyethylene	4
	Soil cement (10:1)	24

9.6. Storage Loss in Pond by Siltation

The above mentioned two losses i.e. evaporation and seepage, reduces the storage capacity of a farm pond. In addition, the storage capacity of a farm pond also gets reduced because of siltation/ sedimentation. The runoff that moves over the soil surface carries some silt/sedimant and when this silt laden water enters to a pond, it gets deposited in it. Sediment load consists of bed load and suspended load. Bed loads are generally bigger in size compared to the suspended loads. Because of gravity, they first get deposited in the bed of the reservoir and decrease the dead storage capacity of the pond. Immediately above the bed, a thin layer consisting of relatively coarser materials with high concentration exists. Above it and extending up to the water surface, a thick layer of relatively fine grained particles exist as suspended load. In case of bed load, movement of particles takes place by rolling, sliding and saltation depending on the velocity of flow. The bed load varies from 10 to 20 per cent of the total suspended load depending on the nature of bed material, its grade, and concentration of suspended materials. The suspended load constitutes the larger share of total sediment load.

9.6.1 Factors Affecting Siltation

Factors affecting the siltation/ sedimentation of farm pond are given as below:

(i) Quantity, quality and concentration of sediment carried by surface runoff from the catchment into the pond
(ii) Percentage of sediments intercepted by the catchment that convey water to the pond
(iii) Percentage of sediments carried away by outflow from the pond through outlet
(iv) Degree of consolidation i.e. weight of silt per unit volume
(v) Length of pond
(vi) Runoff volume and peak discharge from the catchment
(vii) Ratio of pond capacity to annual runoff
(viii) Depth and age of sediment deposited
(ix) Shape and size of pond
(x) Slope of the catchment
(xi) Growth of vegetation at the upper end of the pond
(xii) Soil texture of the catchment area
(xiii) Precipitation and its characteristics in the catchment
(xiv) Temperature and wind velocity of the catchment area
(xv) Geological formation of the catchment

9.6.2 Methods to Reduce Siltation

Siltation decreases the dead as well as live storage of a pond and hence reduces the life span. Proper remedial measures must be taken up to prolong the life span of the pond. The method to reduce the pond sedimentation can be broadly classified into two categories i.e. preventive and curative.

9.6.2.1. Preventive Measures

Preventive measures are better than the curative ones. They are easy and less expensive as compared to the curative measures. If the preventive measures are taken at the right time so that relatively less silt laden water enters into the pond from the catchment, then significant amount of siltation can be checked and pond life span can be enhanced to a tune of 5 to 10 years. The different preventive measures used to reduce the siltation are enumerated as follows (Sharma and Sharma, 2002):

- *Pond site selection*: Farm pond gets silted up due to the sediments coming to it with the runoff from the catchment area. The site where sediment inflow with runoff is high should be abandoned and preferably the catchment producing low sediment should be preferred for construction of pond.

- *Design of farm pond*: In designing a farm pond, ratio of storage capacity to catchment inflow, design and location of outlet works are crucial for estimating the rate of sedimentation. It is recommended that, if economics permit, the life of farm pond should be enhanced by increasing the storage capacity-catchment inflow ratio to such an extent as to have the annual loss of 0.25 per cent only. The ratio between the capacity of the pond and size of the drainage area is one of the most important factors governing the annual rate of sedimentation in the pond. Some amount of sediment flows out with the discharge from the pond through the outlet including sluices. Hence, adequate outlets must be provided in the pond and sluices must be placed at suitable elevations. Proper pond operation can also reduce the sedimentation.

- *Sediment inflow and watershed management*: Sediment inflow to a farm pond can be reduced by constructing check dams on the tributaries bringing highly silt charged waters, constructing contour trenches on the steep slopes of the hills in the catchment area, adopting suitable watershed management practices including biological and engineering measures in the catchment areas of the farm pond and undertaking massive afforestation in the upper side of the catchment areas of the farm pond.

- *Control of sediment deposition*: A farm pond should have sluices and other outlets at different heights. Since the concentration of sediments is not same at all points in the pond, the outlets where the sediment concentration of the pond is high should be operated first so that must of the sediment laden water flows out which ultimately reduces the sediment deposit in the pond.

9.6.2.2. Curative Measures

Silt which is deposited in the pond can be removed by excavation, dredging and sluicing with hydraulic and mechanical agitations. Dredging is too expensive whereas sluicing is accomplished with hydraulic and mechanical agitations for stirring up the silts and then flushing the silt mixed water.

Question Banks

Q1. What are the two measures losses of water in farm pond? How does it vary with respect to the total out flow of water in farm pond?

Q2. How does the evaporation loss of water bodies vary from place to place in India?

Q3. How does the evaporation loss occur in a farm pond?

Q4. How does the open water surface area of a farm pond affect the evaporation rate in it?

Q5. How does the depth of water in a farm pond influence the evaporation rate in it?

Q6. Write the water budget method to estimate the evaporation loss in a farm pond and describe how it is used to compute the evaporation rate.

Q7. Describe the mass transfer and energy budget method to compute the evaporation rate in a farm pond.

Q8. Describe the combination equation for estimation of evaporation loss of farm pond. Write the equation and explain the different terms of the equation.

Q9. Draw the diagram of a USDA Class A Pan evaporimeter and write about its working principle.

Q10. What are the criteria followed in selecting site for a pan evaporimeter?

Q11. What are the different types of evaporimeters?

Q12. An evaporation pan is 1.6 m in size and is used to measure evaporation loss of a farm pond. The initial and final water level reading in the pan are found to be 8 cm and 9 cm, respectively. A rainfall of 4.1 cm has occurred during the period the observation was taken. To keep the water level same in the pan, 2.1 cm depth of water has been removed from the pan. Find out the pan evaporation. If the pan coefficient used to convert the pan evaporation to pond evaporation is 0.75, find out the pond evaporation also.

Q13. Briefly describe the various methods used to reduce the evaporation loss in a farm pond.

Q14. How are the shading materials useful in reducing the evaporation loss in a farm pond?

Q15. What do you mean by biological shading and how are they used to reduce the evaporation loss in farm pond?

Q16. What do you mean by seepage loss in a farm pond? Why it is more in newly constructed ponds?

Q17. How does the seepage loss decreases in the ponds after some years of construction?

Q18. What are the factors that influence the seepage rate in a farm pond?

Q19. How do the depth of water and the wetted surface area in a farm pond affect the seepage rate?

Q20. Write the water balance model of a farm pond and explain how this is used to estimate the seepage rate in a pond.

Q21. Enumerate the various methods employed to reduce seepage loss in farm ponds.

Q22. What is a plastic lined pond? How is the plastic lining helpful in reducing seepage loss in farm pond?

Q23. Write short notes on the followings sealants.
(i) Clay lining (ii) Asphalt lining (iii) Bitumen lining (iv) Monomolecular films

Q24. Write the steps followed for lining a farm pond with plastic films.

Q25. What should be the criteria to select the best sealant for lining a farm pond?

Q26. What do you mean by siltation in farm pond? How does the siltation affect the storage and life span of a farm pond?

Q27. What are the factors that affect the siltation of a farm pond?

Q28. Describe the different preventive and curative measures taken for reducing the siltation of farm pond?

Q29. How does the watershed management planning in the catchment of a farm pond help in reducing the siltation in farm ponds.

References

Bayer , D.G. and Cluff, C.B. 1971. Report of water loss investigation of West Lake, EI, Dorado Country Club. Tucso, Arizona Water Resources Research Centre, Univeristy of Arizona, USA.

Belgaumi, M.I., Itnal, C.J. and Radder, C.D. 1997. Farm Ponds. Technical series 4, Publication Centrte, University of Agricultural Sciences, Dharwad, pp. 47.

Biswas, R.K. 2008. A Few Low Cost Lining Materials. Technical Bulletin, Bidhan Chandra Krishi Viswavidayala, West Bengal, pp. 32.

Biswas, R.K. and Mallick, S. 1997. Performance of locally made burnt clay tiles in controlling seepage losses. J. of Agril. Engg., Vol. 34(4): 47-54.

Bowman, C.C (1967). Sealing Effects of Bentonite and Clay Soils under Extreme Conditions. Agr. Exp. Station Bull., Montana State University, Bozeman, Montana.

Chawla, A.S. 1992. Canal lining and maintenance. Seminar on Irrigation Water Management. Water Management Forum, New Delhi.

Cluff, C.S (1977). The compartmented reservoir, a method of efficient water storage. Ph.D. dissertation, Colorado State University, Fort Collins.

Cooley, K. R.and Idso, S. B.. (1980). "Effects of lily pads on evaporation." Water Resources Research, 16: 605-606.

Cooley, K.R (1974). Evaporation suppression for conserving water supplies. Proceedings of Water Harvesting Symposium, Phoenix, Arizona: 192-199.

Guerra, L.C. Watson, P.G. and Bhuiyan, S.I. 1990. Hydrological analysis of farm reservoirs in rainfed areas. Agricultural Water Management, Vol. 17(4): 351-366.

Hegade, Channappa, T.C. and Anand Ram, B.K. 1981. Water harvesting recycling in the red soils of Karnataka. Ind. J. of Soil Conservation, Vol. 9(2-3): 107-111.

Kamra, S.K. and Dhruvanarayana, B. 1987. Chemical sealant for seepage controlling farm pond. The role of Agricultural Engineering in dry land agriculture. Proceedings of 23rd Convention of ISAE, Yabilpur, India.

Khair, A., Ahmed, M and Data, S.C. 1984. Development of low cost indigenous technology to minimize water losses due to seepage in irrigation canal. AMA, Vol. XV: 77-81.

Nicholaichuk, W. 1978. Seepage control in exacavated earthen reservoirs. Can. Soc. Agril. Engr., Vol. 20(2): 97-101.

Pal, A.R., A.L. Rathore and V.K. Pandey (1994). On-farm rainwater storage systems for improving ricelands productivity in eastern India: opportunities and challenges. In: On-farm Reservoir Systems for Rainfed Ricelandss. Edited by S. I. Bhuiyan. International Rice Research Institute, Manila, Philippines: 105-125.

Panigrahi, B. 2001. Water Balance Simulation for Optimum Design of On-Farm Reservoir in Rainfed Farming System. Ph.D. Thesis, Indian Institute of Technology, Kharagpur.

Penman, H. L. (1948). "Natural evaporation from open water, Bare Soil, and Grass." Proc. Roy. Soc. London, al93, 120-146.

Pochop, L. 0, Borrelli, J. and Hasfurther, V. (1985). "Design Characteristics for Evaporation Ponds in Wyoming." Final *Rep.* submitted to the Wyoming Water Research Center, Laramie, WY, 1-13.

Reginato, R.J., E.S. Nakayam and J.B. Miller (1973). Reducing seepage for stock tank with compacted sodium treated soils. J. Soil Cons., Vol. 28: 214-215.

Rollins, M.B and A.S. Dylla (1970). Bentonite sealing methods compared in the field. J. Irrig. and Drain. Div., (ASCE), Vol. 95(IR2): 193-203.

Shally, H.L. 1985. Lining of Earthen Channel. Asia Publishing House, Bombay.

Sharma, R.K. and Sharma, T.K. 2002. Irrigation Engineering including Hydrology, S. Chand and Company Ltd. New Delhi.

Shen, R.T (1959). Sealing farm ponds and reservoirs with bentonite. Colorado State Univ., Exten. Service Circular 206-A.

Syamsiah, I., Fagi, S. A. M. and Bhuiyan, S. I., 1994. Collecting and conserving rainwater to alleviate drought in rainfed ricelandss of Indonesia. In: On-farm Reservoir Systems for Rainfed Ricelandss. Edited by S. I. Bhuiyan. International Rice Research Institute, Manila, Philippines, pp. 142-152.

U.S. Bureau of Reclamation (1963). Lining for irrigation canal Department of Interior, Washington, 149pp.

Venkatanadhachary, G., Rao, G.P. and Padmaraju, A. 1979. Improved agricultural practices for dryland crops in Rayalaseema region, Andhra Pradesh, Tech. Bull. No. 1, A.P.A.U., Hyderabad.

Verma, H.N., S.S. Prihar and R. Singh (1984). Feasibility of storage of runoff in dugout ponds and its use for supplemental irrigation in submontane Punjab. Indian J. Soil Conserv., Vol. 12(1): 31-36.

Vijayalakshmi, K., K.P.R. Vittal and R.P. Singh (1982). Water harvesting and reuse. In a decade of dryland agriculture research in India (1971-80). AICRPDA, ICAR: 103-119.

Wardana, I.P., Sasa, I.J., Suprapto, M., Fagi, A.M. and Bhuiyan, S.I.. (1996). "On-farm Reservoir System: Management requirements and impact on productivity." Paper presented at the Rainfed Lowland Rice Research Consortium in-house midterm review, 15-19, April, 1996, IRRI, Philippines, 1-18.

Chapter 10

Efficient Use of Harvested Water of Farm Pond

10.1 Introduction

Rainwater is the main source of water for agriculture but its current use efficiency for the crop production is low ranging between 30-45%. The rising demand for water for non-agricultural uses is proportionately decreasing its share for agricultural uses. Efficient methods of water harvesting in the farm reservoirs including farm ponds and its management for subsequent uses in agriculture with improved water use technologies are vital for increasing the productivity and reducing poverty in rainfed ecosystems. It is not to be forgotten that the harvested water in the pond is a very valuable commodity. It may be used for industrial use or for irrigation in agriculture. The harvested water is also a good source to meet the water demand of domestic animals and for human beings (Fig. 10.1). The stored water of the pond can be used for fish farming, tree plantation, intensive irrigation to small area and extensive supplemental irrigation to rainfed crops. Stored water in a pond is developed resource and is, therefore, more expensive than the natural resource. Hence, there is a need to utilize it judiciously and efficiently.

Fig. 10.1 Farm pond for domestic use of animals and human beings

Following conservation of rainwater, its efficient use is achieved through selection of appropriate cropping systems, improved verities, pest and disease management and proper irrigation scheduling methods. Long term on station watershed experiments have demonstrated that vertisols with a rainfall of about 800 mm have the capacity to feed 18 persons per hectare (4.7 tons of food grains/ha) compared with their current productivity of 0.9 tons/ha supporting 4 persons per hectare. This increased productivity can be achieved if productivity of rainwater is enhanced from 30 to 67% and soil loss is reduced by 75% compared with the loss from the traditional methods of cultivation (Wani *et. al.,* 2003).

There are two options to utilize the stored water for irrigated agriculture. In one of the option, the benefit from crop returns is increased by supply of limited supplemental irrigation over large area with small quantity below the water requirement and that to limiting to the critical growth stages of the crop and in the second option, the accrued benefit may be obtained by intensive irrigation according to the optimum water demand over a small area. The former approach allows drought insurance to more area or more farmers and maximum returns of water applied, but this costs more on lengthy water conveyance and application systems, which are to be used for a very small period in year (Zimmerman, 1956). Thus, for a particular area it becomes essential to know whether a fixed amount of irrigation water is utilized more efficiently by the full irrigation to a small area or the limited supplemental irrigation is served to larger area by appropriate water management practice that could otherwise be under rainfed farming. Maximising yield per unit area through intensive irrigation is economically justified where water is readily available and irrigation cost is low, but when water supplies become more limited or irrigation cost is high, the objective of irrigation shall be to maximize yields per unit of available water (Stegmen *et al.,* 1980). The harvested water may be judiciously used for crop life saving (protective) / supplemental irrigation to increase and stabilize crop yield and to increase cropping intensity. The used harvested water should be managed in such a way that stored soil-water and applied water are used by the crop with the highest efficiency. This, of course, requires a comprehensive knowledge on irrigation scheduling of crops.

10.2 Scheduling of Irrigation

To achieve better productivity, it is important to work out an efficient and economic irrigation schedule for water use under any given set of agro-climatic conditions. Irrigation schedules broadly comprises of three steps i.e. (i) when to irrigate (ii) how much to irrigate and (iii) how to irrigate or what method of irrigation is to be adopted for better use of irrigation water in the field.

10.2.1 When to Irrigate?

There are several approaches for scheduling irrigation based on crops, soil, atmosphere and plant water relations. The timing of irrigation is a complex phenomenon depending on a host of factors. It varies from region to region depending on the climatological conditions of the region too. There are several approaches to decide when to irrigate and the most important approaches are described below.

10.2.1.1 Physiological Stages

In this method, irrigations are based on distinct stages of growth. Table 10.1 will help in taking decision at which irrigations are to be given to some of the important field crops based on growth stages after giving the due considerations for the rainfall, especially for *kharif* crops. The real number of irrigations will depend upon agro-climatic conditions, rainfall in crop growth season, ground water contribution and soil types. If irrigation water is not adequately available, then limit the frequency and amount of irrigation at critical growth stages (CGS) only (Table 10.1).

10.2.1.2 Soil Moisture Level

By this method, irrigation is scheduled either on available soil water depletion at certain depth (effective root zone), or soil water potential of the same depths. The magnitude of available soil water depletions at which irrigation is to be scheduled may be a value ranging from 25 to 75 percent while in case of soil water potential, the values may range from 0.1 to 0.7 bars at some specified depth. Table 10.2 shows the available soil water depletion and soil suction at specified depth for some important crops at which water should be applied for maximum yield. Table 10.3 represents the effective rooting depth of some crops and the available soil water of different types of soils (Doorenbos and Pruitt, 1977).

Table 10.1. Physiological stages of crops with respect of irrigations

S. No.	Crops	Physiological stages	Critical stages
1.	Wheat	Crown root initiation (CRI), late tillering, late jointing, flowering, milk and dough	CRI, flowering
2.	Barley	Early vegetative stage, flowering and dough	30 days after sowing, pre-flowering
3.	Potato	Sprouting, stolonisation, stolon development, early tuberisation, 20% of tuber weight, 40% of tuber weight, 60% of tuber weight and 80% tuber weight	Sprouting, early tuberisation
4.	Gram	Pre-flowering and pod formation	Pre-flowering
5.	Soybean	Early seedling, pre-flowering, pod development	Pre-flowering, pod development
6.	Rice	Early tillering, panicle initiation, flowering, milk, dough	Early tillering, flowering
7.	Sorghum	Flowering, dough	Flowering
8.	Maize	Early vegetative growth, tasseling and silking, dough	Tasseling and silking
9.	Cotton	Branching, pre-floweing, boll formation	Pre-flowering, boll formation
10.	Sugarcane	Sprouting, tiller initiation, tillering	Sprouting
11.	Groundnut	Emergence, flowering, pod formation, pod development	Flowering, pod formation
12.	Mustard	Vegetative growth, flowering	Flowering

Table 10.2. Soil water depletion and suctions in the zone of maximum root activity at which water should be applied for maximum yields of some crops

S. No.	Crops	Available soil water depletion (%) at 0-30 cm soil depth	Soil water (suction bar)	Suction at depth, cm
1.	Wheat	50	0.5	30
2.	Potato	25	0.3	30
3.	Sugarbeet	50	0.5	30
4.	Sugarcane	65	0.7	30
5.	Soybean	60	0.6	30
6.	Tabacco	25	0.3	30
7.	Maize	50	0.5	30
8.	Sorghum	75	1.0	60
9.	Cotton	50	0.5	30
10.	Peas (grains)	65	0.7	30
11.	Berseem (fodder)	No depletion	0.3	30

10.2.1.3 Pan Evaporation

Pan evaporation is used to determine the amount of water given in the ratio of irrigation water (IW) and cumulative pan evaporation (CPE). For example if irrigation is scheduled at 1.2 IW-CPE ratios and the depth of irrigation is 6 cm, then irrigation would be given when the CPE reaches 5 cm. The values of IW-CPE ratio varies from crop to crop and depends on climatic condition and ground water table position of the crop field where irrigation by IW-CPE ratio technique is to be adopted. Results of extensive field trials of irrigation scheduling by IW-CPE ratio for various crops by the scientists of All India Coordinated Research Project on Water Management in Chiplima centre of Orissa are compiled and presented in Table 10.4 (Anonymous, 1997). Data of the Table 10.4 also reveals how many irrigations and how much total irrigation has to be supplied by each irrigation schedule to each crop.

Table 10.3. Generalised data on rooting depth of full grown crops, fraction of available soil water, and readily available soil water for different soil types in mm/m depth of soil.

Crop	Effective root depth, m	Fraction of available soil water	Readily available soil water		
			Fine	Medium	Coarse
Alfalfa	1.0– 2.0	0.55	110	75	35
Banana	0.5 - 0.9	0.35	70	50	20
Barley	1.0 - 1.5	0.55	110	75	35
Beans	0.5 – 0.7	0.45	90	65	30
Beets	0.6 – 1.0	0.50	100	70	35
Cabbage	0.4 – 0.5	0.45	90	65	30
Carrots	0.5 – 1.0	0.35	70	50	20
Cotton	1.0 – 1.7	0.65	130	90	15
Cucumber	0.7 – 1.2	0.50	100	70	40
Dates	1.5 – 2.5	0.50	100	70	30
Grains (small)	0.9 – 1.5	0.60	120	80	30
Grains (winter)	1.5 – 2.0	0.60	120	80	40
Grapes	1.0 – 2.0	0.35	70	50	40
Grass	0.5 – 1.5	0.50	100	70	20
Groundnut	0.5 – 1.0	0.40	80	55	30
Lettuce	0. 3 – 0.5	0.30	60	40	25
Maize	1.0 – 1.7	0.60	120	80	20
Melons	1.0 – 1.5	0.35	70	50	30
Olives	1.2 - 1.7	0.65	130	95	25
Onions	0.3 – 0.5	0.25	50	35	45
Peas	0.6 – 1.0	0.35	70	50	40
Potatoes	0.4 - 0.6	0.25	50	30	30
Safflower	1.0 – 2.0	0.60	120	80	15
Sorghum	1.0 – 2.0	0.55	110	75	50
Soybeans	0. 6 – 1.3	0.50	100	75	35
Spinach	0.3 - 0.5	0.20	40	30	35
Strawberries	0.2 – 0.3	0.15	30	20	15
Sugar beet	0.7 – 1.2	0.50	100	70	10
Sugarcane	1.2 – 2.0	0.65	130	90	30
Sunflower	0.8 – 1.5	0.45	90	60	40
Sweet potato	1.0 – 1.5	0.65	130	90	30
Tomatoes	0.7 – 1.5	0.40	180	60	40
Vegetables	0.3 – 0.6	0.20	40	30	25
Wheat	1.0 – 1.5	0.55	105	70	15

N.B. The above mentioned values of fraction of available soil water is meant when ETc = 5-6 mm/day. When ETc is 3 mm/day or less, increase the value by 30% and when ETc = 8 mm/day or more, decrease the value by 30% (Doorenbos and Pruitt, 1977).

Table 10.4 Values of IW: CPE at which irrigation is scheduled and irrigation requirement of some crops in Chiplima, Orissa

Crop	Duration (days)	IW:CPE	Water required at each irrigation (cm)	Number of irrigation	Total amount of irrigation (cm)
Ground nut (Summer)	115	1.2	6	8	48
Mustard	75	0.9	6	3	18
Potato	90	1.2	5	4	20
Sugarcane	360	1.2	6	15	90
Wheat	95	1.0	6	5	30
Green gram	60	0.8	6	3	18
Black gram	60	0.8	6	3	18
Sesamum (Summer)	90	0.9	6	6	36
Sunflower	90	1.0	6	7	42
Maize	90	0.9	6	5	30
Chilli	150	1.0	6	6	36
Onion	105	1.2	5	8	40

Irrigation to rice is recommended to be given at 3 days after disappearance of ponded water (ADP) in *kharif* and *for rabi* rice, irrigation is to be given at 1 ADP and during each irrigation 7 cm water may be supplied (Anonymous, 1997). It is important to remember that the above mentioned IW:CPE irrigation schedule suggested for various crops is to be followed when there is no deficit of irrigation water. However, if there is dearth of supply form the pond, we have to choose to supply water to the crop at the most responsive stage i.e. critical growth stage which is presented in Table 10.1 for some important crops.

10.2.1.4 Irrigation Scheduling based on Calculations

Irrigation interval is calculated by the formula given below:

$$\text{Irrigation interval} = \frac{\text{Allowable soil water depletion}}{\text{daily water use}} \tag{10.1}$$

For example in case of wheat grown in loam soil, available soil water is 10 cm (effective root zone depth = 60 cm) and allowable soil water depletion is 50 percent. Daily water use is 0.5 cm. Allowable soil water depletion is calculated as 10 x 0.5 = 5 cm. Since daily water use is 0.5 cm, the irrigation interval is computed as 5/0.5 = 10 days. Therefore, irrigation should be given at 10 days interval.

Sample Calculation

Calculate the irrigation interval for potato grown in Chiplima with the help of the following data.

Field capacity, F C = 22.0% dry weight basis

Permanent wilting point, PWP = 10% dry weight basis

Bulk density, BD = 1.40 g/cc

Effective root zone depth, RD = 60 cm

Daily water use, ETcrop = 0.25 cm/day

Allowable soil water depletion = 25 percent of available soil water.

Solution

$$\text{Available soil water} = \frac{(FC - PWP)}{100} \times BD \times RD$$

$$\text{So, available water} = \frac{(22.0 - 10.0) \times 1.40}{100} \times 60 = 10.08 \text{ cm}$$

$$\text{Allowable soil-water depletion} = \frac{25 \times 10.08}{100} = 2.52 \text{ cm}$$

$$\text{Interval of irrigation} = \frac{2.52}{0.25} = 10.08 = 10 \text{ days}$$

Sample Calculation

After how many days will you irrigate a tomato crop if?

Field capacity of the soil = 26% (dry weight basis)

Permanent wilting point = 11% (dry weight basis)

Bulk density of soil = 1.3 g/cc

Effective depth of the root zone = 70 cm

Daily consumptive use of water for the crop = 10 mm

Allowable soil water depletion = 100 percent of available soil water

Solution

Available depth of water or deficit d in the root zone depth is :

$$\text{Available soil water} = \frac{FC - PWP}{100} \times BD \times RD$$

where

FC = field capacity = 26%
PWP = permanent wilting point =11%
BD = Apparent specific gravity =1.3 g/cc
RD = Effective depth of the root zone = 70 cm

$$\text{Available soil water} = \frac{(26-11)}{100} \times 1.3 \times 70 = 13.65 \text{ cm}$$

$$\text{Allowable soil-water depletion} = \frac{100 \times 13.65}{100} = 13.65 \text{ cm}$$

Therefore, number of days between watering = $\frac{13.65}{1.0}$ = 13.65 days (14 days say).

10.2.2. How Much to Irrigate? (Net Quantity of Water to be Applied)

After deciding when to irrigate a crop, next question comes how much water to be irrigated to a crop at each irrigation. The net quantity of water to be applied depends on magnitude of moisture deficit in the soil, leaching requirement and expectancy of rainfall. When no rainfall is likely to be received and soil is not saline, net quantity of water to be applied is equal to the moisture deficit in the soil, i.e. the quantity required to fill the root zone to field capacity. The moisture deficit (d) in the effective root zone that is to be filled up by irrigation is found out by determining the field capacity moisture content and bulk density of each layer (Misra and Ahmed, 1987).

$$d = \sum_{i=1}^{n} \frac{(FCi - PWi)}{100} \times Asi \times Di \qquad \text{... (10.2)}$$

where, $\sum_{i=1}^{n}$ = summation of n-number of layers in the root zone;

FCi = field capacity of the ith layer on oven dry weight basis; PWi = actual moisture content of the ith layer on oven dry weight basis; ASi = apparent specific gravity of ith layer, and Di = Depth of ith layer.

Sample Calculation

Find out the net quantity of irrigation water to be applied to ground nut with the following values of soil moisture content and other data.

Depth of soil layer in (cm)	Moisture % on oven-dry basis		Apparent specific gravity, g/cc
	Field capacity	Actual	
0-15	26.0	17.4	1.39
15-30	24.0	17.8	1.47
30-60	22.3	19.2	1.51
60-90	22.2	20.5	1.53

Solutions

Moisture deficits in the different layers will be as follows:

$$\text{1st layer} = \frac{26.0-17.4}{100} \times 1.39 \times 15 = 1.79 \text{ cm}$$

$$\text{2nd layer} = \frac{24.0-17.8}{100} \times 1.47 \times 15 = 1.36 \text{ cm}$$

$$\text{3rd layer} = \frac{22.3-19.2}{100} \times 1.51 \times 30 = 1.40 \text{ cm}$$

$$\text{4th layer} = \frac{22.2-20.5}{100} \times 1.53 \times 30 = 0.78 \text{ cm}$$

So, net quantity of water to be applied is 5.33 cm.

10.2.2.1 Gross Irrigation Requirement

The total amount of water applied through irrigation is termed as gross irrigation requirement. In other words, it is net irrigation requirement plus losses in water application and other losses. The gross irrigation requirement can be determined for a field, for a farm, for an outlet command area or for an irrigation project, depending on the need, by considering the appropriate losses at various stages of water conveyance and distribution.

$$\text{Gross irrigation requirement (in field)} = \frac{\text{Net irrigation requirement}}{\text{Field efficiency of the system}} \quad ...(10.3)$$

Sample Calculation

If net amount of irrigation is 5 cm and the field efficiency is 50 per cent, then find out the gross amount of water to be applied to the field.

Solutions

Net irrigation requirement = 5 cm

Field efficiency of the system = 50 per cent

Hence, gross irrigation requirement = 5 / 0.50 = 10 cm.

10.2.2.2. Seasonal Gross Irrigation Requirement

The seasonal gross irrigation requirement (IR) at the field head can be determined as follows:

$$IR = \frac{\sum_{i=1}^{n} d_i}{E_a} \quad ... (10.4)$$

where, IR = seasonal gross irrigation requirement at the field head, cm; d = net amount of water to be applied to bring the soil moisture level in the effective root zone of the crop to field capacity, cm; Ea = water application efficiency (in decimal), and n = number of irrigations in a season.

10.2.3 Irrigation Water Measurement

The volume of water supplied to a given area in unit time is called as discharge. The discharge measurement in agricultural operations is very important to ensure economic and optimum utilization of irrigation water. It also helps in better utilization of irrigation water and is vital in irrigation water management. The methods of discharge measurement can be broadly classified into the following types: (i) volume method of discharge measurement (ii) velocity-area methods (iii) slope-area method (iv) flow measuring structures such as orifices, flumes and weirs and (v) chemical dilution method. Out of these methods, the volume method is less costly and simple. The irrigation water measurement through the flumes, especially Parshall and cutthroat flumes are mostly adopted in the fields.

10.2.3.1. Volume Method of Discharge Measurement

This is a simple method for discharge measurement used in irrigation channels. In this method, the flow of water of the stream or the small irrigation channel in a container such as bucket of known volume for a given time period is collected (Fig. 10.2). The rate of flow/ discharge of the stream/small irrigation channel is computed by the formula:

$$\text{Discharge, lit/sec} = \frac{\text{Volume of container, lit}}{\text{Time required to fill the container, sec}} \quad \ldots (10.5)$$

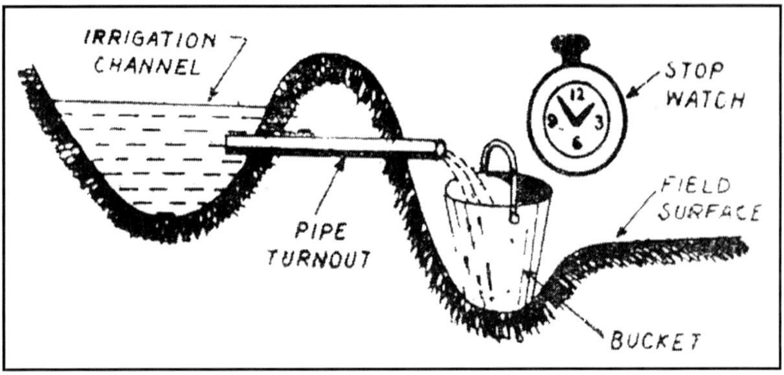

Fig. 10.2 Installation for volume method of discharge measurement

Sample Calculation

A 20 lit capacity container is filled with water flowing in an irrigation channel in 5 sec. What is the discharge flowing in the channel?

Solution

Volume of the container = 20 lit.

Time require to fill the container with water of the channel = 5 sec.

$$\text{Hence, discharge} = \frac{20 \text{ lit}}{5 \text{ sec}} = 4 \frac{\text{lit}}{\text{sec}}$$

Volume method can be used to determine discharge of pumps and other water lifting device such as Persian wheel, chain washer pump, indigenous water lifting devices such as tenda, doon, swing lift, wheel lift, wind mill etc. The irrigation water to a crop field can also be measured by a pump. By knowing the volume of water required by a crop and the pump discharge capacity, we can find out what a time the pump has to run to irrigate the field. In other words, the discharge capacity of a pump which is irrigating a crop field can also be calculated by volume method by knowing the irrigated area, depth of water irrigated and the time required to irrigate the filed by the pump. Multiplying the irrigated area with depth of irrigation, volume of water is found out. Then dividing the volume of water with the time required to irrigate the field, discharge capacity of the pump is found out. A few important pumps generally used to irrigate the fields are (i) low lift hand pump (ii) centrifugal pump (iii) pedal operated pump (iv) krishak bandhu pump. In addition, many farmers irrigate their crop fields from the farm pond by indigenous water lifting devices such as tenda, doon, swing lift, wheel lift, wind mill etc. Details of various pumps and indigenous water lifting devices used for irrigation in fields are discussed subsequently in this chapter.

Sample Calculation

A pump is irrigating a field of 1 ha and the average depth of irrigation water is 5 cm. If the time required for irrigation is 5 hr, then find out the discharge of the pump.

Solution

Volume of irrigation water = Area x average depth of irrigation = 10000 m^2 x 0.05 m = 500 m^3.

Time of irrigation = 5 hr

So, discharge of pump = 500 m^3 /5 hr = 100 m^3/hr = 27.78 lit/sec.

Sample Calculation

Wheat crop require 40 cm of irrigation water during 120 days irrigating period. If area under cultivation is 10 ha and the pump used for irrigation is engaged every day for 10 hours each, then find out the discharge of the pump.

Solution

Total volume of water during the irrigation period of 120 days = 0.40 m x 10 x 10000 m^2 = 40, 000 m^3.

Total time of irrigation in 120 days @10 hours per day = 120 x 10 = 1200 hr.

Discharge of the pump = Total volume of water/ total time of irrigation

= 40,000 /1200 = 33.33 m3/hr = 9.26 lit/sec.

10.2.3.2. Parshall Flume

Parshall flumes consists of (i) a converging inlet section (ii) a throat section and (iii) a diverging outlet section. The throat section lies in between the inlet and outlet sections. Fig, 10.3 shows the geometry of a Parshall flume. Parshall flumes may be built of wood, concrete, galvanized sheet metal or other materials. Large flumes are usually constructed on site, while smaller ones can be purchased as prefabricated units that are installed in the channel. Dimensions and ranges of flow for the various standard Parshall flumes as well as the design procedures can be found in Water Measurement Manual of the U.S. Bureau of Reclamation (1975).

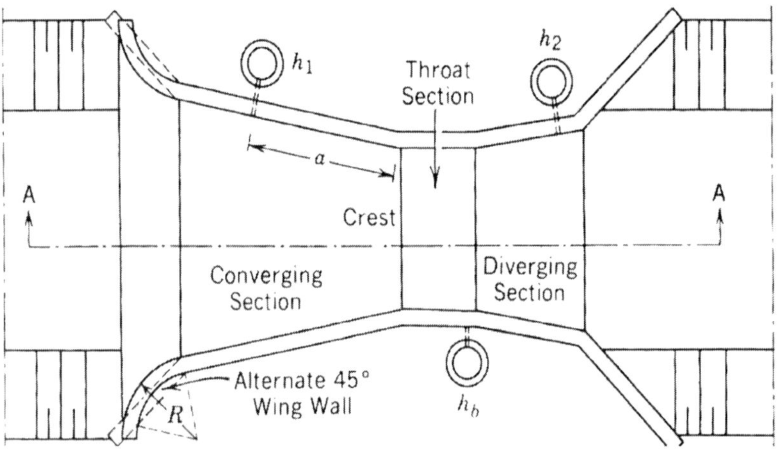

Fig. 10.3. Geometry of a Parshall flume

The primary advantage of Parshall flumes is their ability to provide accurate flow measurements over a wide range of flows with minimum of head loss (U.S. Bureau of Reclamation, 1975). This allows their uses in relatively shallow channels with flat grades. The main disadvantages of Parshall flumes are their relatively large size (Walker and Skogerboe, 1987).

The rate of flow through a Parshall flume may or may not be affected by the tail water depth in the downstream of the throat. Free flow condition exists when the tail water depth is not high enough to affect the flow. A flume is considered to be submerged when the tail water depth is sufficient to affect the flow and the submergence ratio, S defined as the ratio of head in the diverging section (H_b) to that in the converging section (H_a) exceeds the transition ratio S_t. Values of S_t for different sizes of Parshall flumes as decided by the throat widths are presented in Table 10.5 (James, 1988).

Table 10.5. Free flow and submerged flow coefficients and exponents for Parshall flume

Throat width, W, cm	C_f for Q in (lit/s)	C_s for Q in (lit/s)	n_f	n_s	S_t
2.54	9.57	8.47	1.55	1.00	0.56
5.08	19.14	17.33	1.55	1.00	0.61
7.62	28.09	25.91	1.55	1.00	0.64
15.24	58.34	47.01	1.58	1.00	0.55
22.86	86.94	71.08	1.53	1.08	0.63
30.48	113.28	88.08	1.52	1.06	0.62
45.72	169.92	125.17	1.54	1.08	0.64
60.96	226.56	168.22	1.55	1.115	0.66
76.20	283.2	204.47	1.555	1.14	0.67
91.44	339.84	243.55	1.56	1.15	0.68
121.92	453.12	314.35	1.57	1.16	0.70
152.40	566.40	383.74	1.58	1.185	0.72
182.88	679.68	448.87	1.59	1.205	0.74
213.36	792.96	514.01	1.60	1.230	0.76
243.84	906.24	577.73	1.60	1.250	0.78
304.8	1136.48	702.05	1.59	1.260	0.80

The following head discharge equation is used for free flow Parshall flumes (James, 1988).

$$Q = C_f \left(K H_a \right)^{n_f} \quad \text{... (10.6)}$$

where, Q = discharge, lit/sec; C_f = Free flow coefficient depending on throat width (Table 10.5.); n_f = free flow exponent (Table 10.5); H_a = head measured in the converging section of the flume, m and K = unit constant taken as 3.28 when H_a is in m.

In submerged case, the ratio of head in the diverging section (Hb) and that in the converging section (H_a) i.e $S = H_b/H_a$ > transition submergence, St. Table 10.5 can be referred for deciding submergence case considering the value of St. In submergence case, discharge is given (James, 1988) as:

$$Q = \frac{C_s \left[K\left(H_a - H_b\right)\right]^{n_f}}{\left[-\left(\log S + C\right)\right]^{n_s}} \qquad \text{... (10.7)}$$

where, Q = discharge, lit/sec; C_s = submerged flow coefficient depending on throat width Table 10.5); n_f and n_s = free and submerged flow exponent (Table 10.5); H_a and H_b = heads measured in the converging and diverging sections of the flume, m, respectively and K = unit constant taken as 3.28 when H_a is in m, S = ratio H_b to H_a, and C = 0.0044.

Sample Calculation

A Parshall flume of throat width 7.5 cm is used to measure discharge. If the head measured in the converging section of the flume is 30 cm and the flow is under free flow, find out the discharge measured by it.

Solution

Given throat width, W = 7.5 cm; H_a = 30 cm = 0.3 m and free flow condition.

Referring Table 10.5 for W = 7.5 cm, C_f = 28.09; n_f = 1.55.

Using H_a = 0.30 m and K = 3.28 (when H_a is in m),

Discharge is obtained as

Q = 28.09 x (3.28 x 0.3) 1.55 = 27.4 lit/sec.

Sample Calculation

The discharge is measured by a Parshall flume having the following data.

Throat width, W = 7.5 cm; head measured in the converging and diverging section of the flume are 45 and 30 cm, respectively. Find out the discharge measured by the flume.

Solution

Given, throat width, W = 7.5 cm; H_a = 45 cm = 0.45 m, H_b = 0.35 m.

Value of S = H_b/H_a = 0.30/0.45 = 0.67.

Referring Table (10.5) for W = 7.5 cm, we find S_t = 0.64

Since S > St, it is submerged case.

Now refereeing Table (10.5), for W = 7.5 cm and under submerged case, C_s = 25.91; n_f = 1.55, n_s = 1.00; C = 0.0044; S = 0.67; K = 3.28 (H_a is taken in m).

Using the above data and Eq. (10.7), discharge is obtained as = 50.90 lit/sec.

10.2.3.3. Cutthroat Flume

A cutthroat flume consists of (i) level floor (ii) a uniformly converging inlet section (iii) a uniformly diverging outlet section and (iv) a throat occurring at the intersection of the inlet and outlet sections. A plane view of a cutthroat flume and construction details are shown in Fig. 10.4.

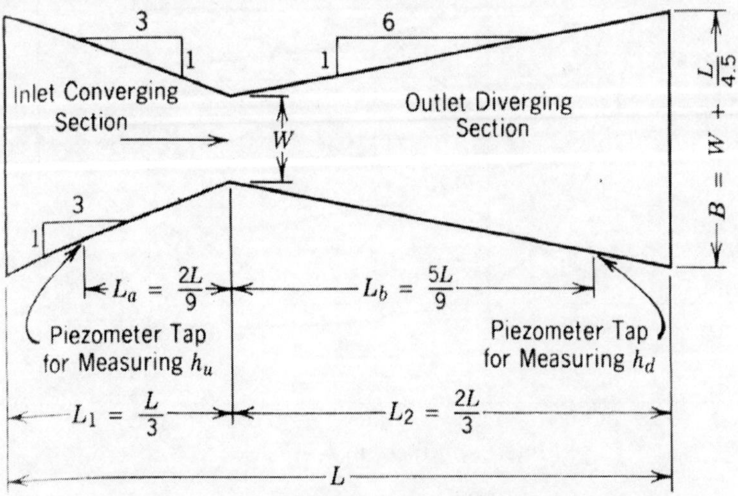

Fig. 10.4 Geometry of a cutthroat flume

Cutthroat flumes are easier and more economical to build and install than the Parshall flume. The same equations as used for discharge determination for free and submerged flow conditions in Parshall flume (Eqns. 10.6 and 10.7, respectively) are also used for discharge computation in cutthroat flume for free and submerged flow conditions, respectively. However, the parameters of these equations like C_f, C_s, n_f, n_s, S_t, K_f and K_s are different and are given in Fig. 10.5 (James, 1988). Value of C is considered as zero. The following equations are used to calculate C_f and C_s.

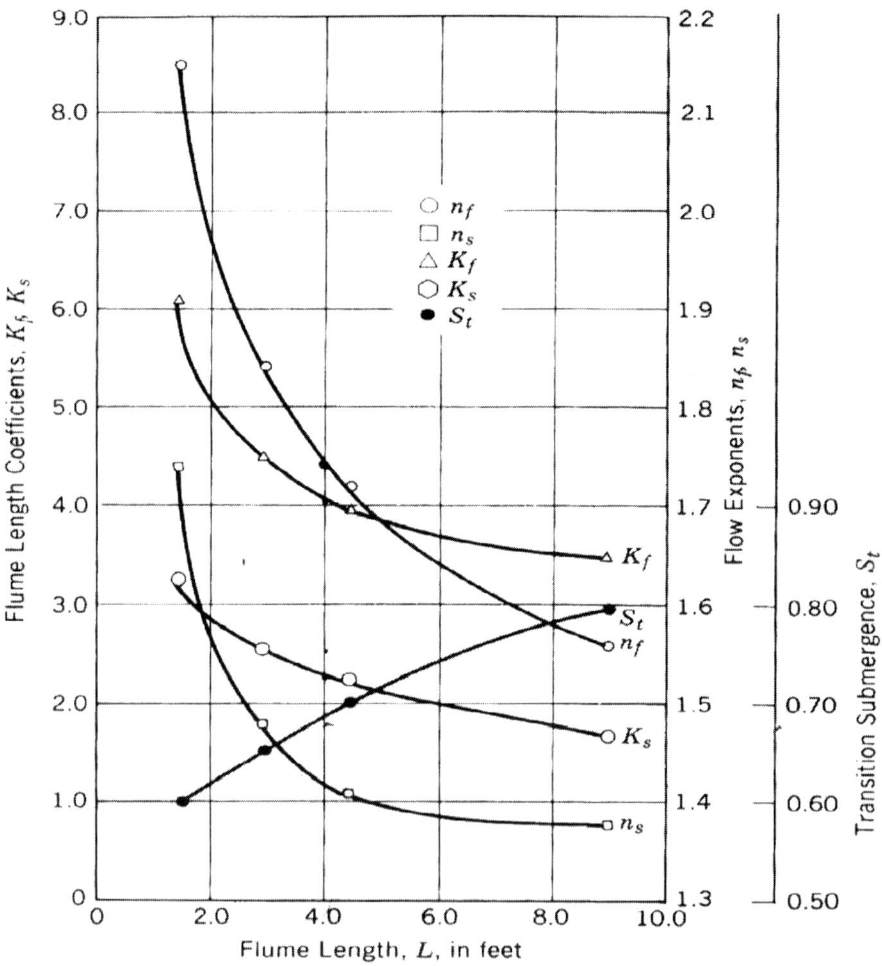

Fig. 10.5. Free flow and submerged flow coefficients (K_f and K_s), exponents (n_f and n_s) and transition submergence (S_t) for cutthroat flumes

$$C_f = K K_f W^{1.025} \quad \text{... (10.8)}$$

$$C_s = K K_s W^{1.025} \quad \text{... (10.9)}$$

where, W = throat width, ft; K_f and K_s are constants for free and submerged flow conditions, respectively (taken from Fig.10.5); K = unit constant (1 for Q and H in Eqns. 10.6 and 10.7 in ft^3/sec and ft, respectively and also for H_a and H_b in Eq. 10.7 in ft).

Sample Calculation

A cut throat flume is used to measure discharge under free flow condition for irrigating ground nut field. The flume has a length 4 ft. The throat width of the flume is 3 ft. The head measured in the converging inlet section of the flume is 2 ft.

Solution

Given length of flume, L = 4 ft.

Width at throat, W = 3 ft.

Head at the converging inlet section of the flume, H_a = 2 ft.

From Fig. 10.5, when length, L = 4 ft, n_f = 1.74 and K_f = 4.1.

From Eq. (10.8),

Putting K = 1, K_f = 4.1 and W = 3, C_f is found as = 12.64

Under free flow case, discharge $Q = C_f [K \times H_a]^{n_f}$ (Eq. 10.6)

Putting C_f = 12.64, K = 1, H_a = 2 and n_f = 1.74,

Discharge is obtained as = 42.23 ft^3/sec which is equivalent to 1195.54 lit/sec.

Sample Calculation

Find the discharge measured by a cutthroat flume installed in a channel with the following recorded data.

Length of flume = 4 ft; width at throat section = 3 ft; upstream head = 2 ft; downstream head = 1.6 ft.

Solution

Given length of flume, L = 4 ft.

Width at throat section = 3 ft.

Upstream head, H_a = 2 ft and downstream head, H_b = 1.6 ft.

Submergence ratio, S = H_b/ H_a = 1.6/2.0 = 0.8

From Fig. 10.5 at length of flume, L = 4 ft, transition submergence, St is 0.68, constant K_s is 2.3, n_f is 1.74 and n_s is 1.42.

Since S > St, flume is under submergence case and so discharge formula under submergence case i.e. Eq. 10.7 is to be used.

From Eq. (10.9), $C_s = K K_s W^{1.025}$

Putting K = 1, K_s = 2.3, and W= 3, C_s is 7.092

Under submerged case discharge, Q is given by Eq. 10.7.

Putting Cs= 7.092, K = 1, C = 0, n_f = 1.74, n_s = 1.42, S = 0.8, H_b = 2 ft and H_a = 1.6 ft in Eq. 10.7, discharge is obtained as 39.60 ft^3/sec which is equivalent to 1121.11 lit/sec.

10.2.4 Methods of Irrigation

The third criterion of irrigation schedule is how to irrigate a crop when we have decided when to irrigate and how much quantity of irrigation water is to be given to the crop. This is adjudged by methods of irrigation. When water flows from the canal to the field a portion is soaked into soils in all directions. This is called seepage. When areas near by are saturated water moves downward into deeper layer. This is called deep percolation. Water lost due to seepage, deep percolation and evaporation does not help to meet water requirement of crops. These losses to a greater extent are avoidable losses. Mechanical composition of soil and length, breadth and size of the canal and the slope of the land, in general, affect the seepage and deep percolation losses. Such losses in sandy soils are more than that in clayey soils. Reduced flow velocity in long and flat canals increases seepage and deep percolation losses. Growth of weeds and phreatophytes obstructs the free flow and encourages such losses too.

Depending on source of water, type of crop to be irrigated, quantity and quality of water to be applied and other economic considerations, irrigation water may be applied to crops from surface, sub-surface, under pressure (sprinkler) or in drops (drip irrigation). The most popular and easy form of irrigation is the surface irrigation.

10.2.4.1. Surface Irrigation

Surface irrigation method is most widely practiced. In this system water is conveyed to the point of infiltration directly on to the soil surface in channels that vary in shape, size and hydraulic characteristics. The methods may vary from corrugation to long narrow strips or large fields where water is impounded. The most widely used surface irrigation method is done by free flooding. The method is easy and inexpensive. Crops like rice, wheat, oilseeds and fodder crops and other close growing crops are generally irrigated by this method (Fig. 10.6). This method is also called wild flooding. In this method water flows from the ditch directly to the field without much control on either side of the flow. It covers the entire field and moves almost unguided. The rate of advancing front is controlled by the topography of the field. Uneven distribution of water and low water application efficiency are the common drawback of this method.

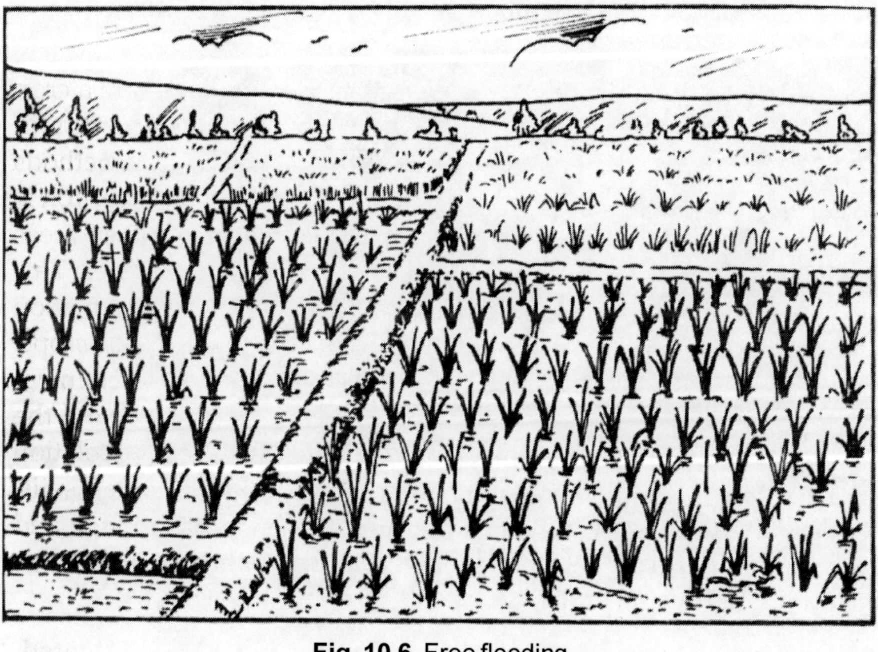

Fig. 10.6. Free flooding

In addition to free flooding, there are other surface irrigation methods like border strip, furrow, check basin, basin or ring etc. Different crops need different methods of irrigation and each method has some limitations (Table 10.6).

Table 10.6. Methods of irrigation for different crops and their limitations

Method	Crops to be irrigated	Topography	Stream size	Remarks
Free flooding	Rice, wheat, broadcast crop, fodder, millets	Sloppy and irregular up to 10%	Large	Inefficient method, causes unequal distribution of water
Boarder strips	Wheat, rice, sorghum, pulses	No cross slope, slope along length	Small to medium	Sandy loam to loam. graded, low layout cost, easy for intercultural works
Furrow (ridge and furrow)	Sugarcane, potato, cotton, Vegetables	Flat and gentle slopes.	Small to large supply	Economic water use , facilitates intercultural works
Corrugated method	Vegetables, onion garlic, spinach	Slopes perpendicular to furrow	Small to large supply	Economic water supply, interculture becomes difficult
Broad ridge and narrow furrow	Turmeric, ginger, yam, sweet potato	-	8-12m long, breadth depends on crops	Almost flat gentle slope in both directions
Basin or ring	Orchard crops, cucurbits	Flat	1-3 m size ring, 40-45 cm deep	Economic water use increases efficiency.
Check basin	Grain crops orchard crops	-	Small to medium	Low maintenance cost; erosion control

Now–a-days micro irrigation methods including sprinkler and drip are used to irrigate the crops which require low amount of irrigation water and gives high yield and water use efficiency. Micro irrigation is especially useful for vegetables, flowers and other horticultural crops including high value crops.

10.2.4.2. Sprinkler Irrigation

By this method water can be supplied at rates equals to, lesser than or greater than the intake rate of the soil. Sprinkler irrigation is used for almost all crops and or nearly all soils. it may be the only way that crops can be irrigated satisfactory on soils having a very high rate intake, steep slopes, shallow depth or irregular topography where no other system is feasible. In sloppy land, levelling and grading involves cut and fill of the soils, disturbs top soil and brings heterogeneity in soil fertility and productivity. Sprinkler systems in such a soils become more economic without loss of land for the bunds, channels and ditches. Sprinkler system works the best if stream size is small and quantity of water is available is small. Time and technique required for efficient execution of the other methods of irrigation can be saved if sprinkler irrigation method is followed. High degree of water control is the most important advantages of sprinkler system. Water application rates can be as low as 0.25 cm/hr without the hazards of the soil erosion or excessive water losses in the sallow and highly permeable soils of low water holding capacity. Besides meeting the water requirement of crops, sprinkler can also be an effective method of supplying nutrients through foliage. It can also combine insecticides and pesticides. It can control and protect the crop from frost damage and temperature stress. Frosting damages almond, arrowroot, straw berry, blue berry, cranberry, citrus fruits and flowers of different types and sprinkler irrigation reduces these damages. It also served the fruit tree branches from breaking due to heavy load of frost deposits. Crops like tea and coffee grown in the plain and hill slops effectively irrigated by this method. Shallow rooted crops like onion, garlic, tomato, potato and lawns can most effectively be irrigated by sprinkler irrigation. In some cases if due to desiccation of top soil germination and emergence of seedling are affected and other method of irrigation cannot supplied the required small quantity of water, sprinkler method proves the most efficient one. It eliminates surface crusting and impounding of water leading to seedling rotting.

Sprinkler system has disadvantages too. Sprinkler low quality irrigation water damages the citrus crops. In some cases it leaves undesirable deposits on leaves and fruits. Free water on the crop surfaces induces growth of fungi and bacteria and multiplies diseases. Sprinklers closed to the fruits like apple and grapes cause bruises on their surface and reduce quality. Small, slender and tender seedlings break due to sprinkler force.

Before adopting the sprinkler system advantages and disadvantages of different systems need to compare from above point of views beside the high initial cost involves. One sprinkler set costs about Rs. 30,000-40,000 under Indian conditions which is not within easy reach of an ordinary farmer. It proves economic in industrial crops like tea, coffee, pepper, cardamom and fruit orchards. Common application of sprinkler system in general agriculture has not yet been found economic in our country. In arid regions of India, sprinkler system is gradually gaining popularity. In these regions of cropping pattern consists of the low water requirement crops like wheat, mustard, gram and adoption of sprinkler irrigation system will increase their water use efficiency.

The essential parts of a sprinkler system are: (1) a hydrant or water pump, (2) large aluminum pipe, (3) narrow aluminum pipe, (4) risers and, (5) sprinkler nozzles. The main pipe is coupled with the hydrant or the water supply pump; to the main pipe are attached the laterals and to the laterals the risers. At the top, the risers have the nozzles fitted. The laterals are fitted to the main pipes at 15-20 m distance either on one side or both the sides. Water supplied to main by force gets distributed to the nozzles that operate by trigger mechanism.

On the basis of portability of sprinkler units, these are classified as: (i) portable (ii) semi-portable and (iii) stationary sprinkler. Portable and semi-portable types are most commonly used in agriculture. This system can be moved manually or mechanically.

On the basis of nozzle types, sprinkler may be fixed nozzle type, perforated type and rotating type. In the fixed nozzle type, the sprinklers are fixed in a row parallel and water sprinklers out at right angle from the pipes. By changing the pipes the plots are irrigated. In perforated pipe system, there are perforation in the lateral irrigation pipes in a specially design pattern to distribute water almost uniformly and the system works with low operating pressure of about 0.5-2.5 kg/square cm. the sprays can be on either side of pipe.

In rotating head system, nozzles are placed in risers fixed at uniform interval along the length of lateral pipe (Fig. 10.7). Crops like tomato, potato, sugarbeet and many flowering plants are best irrigated by this. Soil with low in take rate and high water holding capacity is generally irrigated by rotating sprinkler systems. Most agricultural sprinklers are rotating types with either one or two nozzles. That varies in size from 1.5 mm to 5 cm. the discharge from such nozzles varies from 4 liters to 4000 liters per minute and covers circular area from 6 m o 180 m in diameter. The most commonly used ones are with capacities ranging from 7.5 to 75 liters/minute that operate at pressure from 1.4 to 4.2 kg/square cm. The rotation is caused by a torque produced by a reaction of water leaving the nozzle or due to impact of a spring loaded arm. Some sprinkler is gear driven by a small turbine located at the base of the sprinkler or externally on ones side of the nozzles.

Fig. 10.7. Rotating head sprinkler in operation

The soil puddles and runoff begins if the drop becomes very large. Not to affect the crop soil the drops should be small that depend on nozzle size and pressure. For small size particles with nozzle size of 3-5, 5-6 and 6-9 mm the pressure respectively should be 6-9, 8-11 and 9-13 kg/ square cm. Water distribution depends on wind velocity and is affected if the velocity become more than 16-17 km/hr. If wind velocity is high, the lateral and risers spacing should be closer.

Uniformity of Distribution

The basis objective of sprinkler irrigation is to supply water to the crop just like rainfall and meet the crop water requirement with maximum water utilization efficiency, this deepens upon how much water to be used, at what rate to sprinkler and how best to distribute so that it will be uniform throughout the crop. The uniformity of distribution is regulated by the spacing of sprinkler, type of sprinkler nozzle and operating pressure. Low pressure result in large drops sprinkled away from the sprinkler and high pressure, on the other hand, breaks the water particles into smaller units and sprinklers very near to sprinkler. Wind velocity distorts the distribution pattern. Uniformity of distribution is expressed by distribution efficiency which is defined as:

$$E_d = (1 - \frac{\bar{y}}{\bar{d}}) \times 100 \qquad \text{... (10.10)}$$

where, E_d = distribution efficiency; $\bar{d}$ = average depth of water stored along the run during irrigation and $\bar{y}$ = average numerical deviation from $\bar{d}$.

In case of sprinkler irrigation, uniformity is evaluated by calculating uniformity coefficient as:

$$Cu = \left(\frac{1 - |X|}{M \cdot N} \right) \cdot 100 \qquad \text{... (10.11)}$$

where, X = absolute value of deviation of the individual observation of the depth applied from the mean value M; N is number of equally spaced observations of depth applied for the sprinkler pattern.

Data on uniformity coefficient are useful to select suitable combination of spacing, discharge, nozzle size, and operating pressure to get high irrigation efficiency.

10.2.4.3 Trickle Irrigation (Drip Irrigation)

Drip is one of the latest methods of irrigation and is more specified one. It is increasingly becoming popular, particularly in water scarcity areas. In this method water is supplied in continuous or discontinuous small drop to the root zone of the crop. entire cropped area, unlike in other methods of irrigation including sprinkler is not required to be wetted.

It irrigates the crop frequently but at rate equivalent to its consumptive use criminating conveyance, seepage and deep percolation losses and this result in the maximum efficiency. Moisture regime remains around field capacity and crop is not stressed. It can also irrigate with poor quality water without building up soil salinity to hazardous level. This method reduces pests, salinity hazard, operational labour and energy requirement. The system is more efficient and more specialized than sprinkler irrigation system. high cost of installation makes it prohibitive for the general use. It, however, can be profitable used in orchards, vegetable fields in water scarcity arid zones. Emitter clogging, rodent or animal damage, inadequate soil water movement, plant root development and operational technical difficulties are some of the disadvantages of this method.

The trickle irrigation system consists of a main line, submain, laterals and ticklers. Like in sprinkler system, the main fed by a pump feeds the submain, the submain the lateralas and the laterals to the tricklers. These components are made of polyvinyl chloride (PVC) tubing which can withstand saline irrigation water and fertilizer action. The ancillary components are valve, pressure gauge and filters. For designing trickler system knowledge on consumptive use of crop. Its pacing, slope and topography of land, water intake rate, etc. are essential.

A pump of suitable capacity supplied required pressure to force water through main unto emitter. The laterals may operate with pressures as low as 0.15-0.2 kg/square cm and as high as a 1 to 1.75 kg/square cm; the pressure gradually drops at the laterals and emitters is almost at the atmospheric pressure. If desirable a fertilizer tank may be connected with the system to supply different nutrients to the crop with irrigation. The drip nozzles are provided at regular intervals on the laterals. The emitters allow the water to trickle at very low rates. Fig. 10.8 shows the layout of the drip irrigation system in crop field.

Fig. 10.8. Layout of drip irrigation system

10.2.5. Water Lifting Devices

Unlike canal irrigation where water is fed to the fields through gravity system, water from the farm pond is lifted either mechanically by pump or manually by any indigenous lifting devices. Irrigation by lifting water from a lower level to a higher level is called as lift irrigation. Several types of indigenous water lifting devices are in vogue in India since ancient times. These are animal or/and manually operated ones, the animal operated ones being more efficient than the manually operated ones. Swing basket, Sedua, Doon, Tenda, Wheel lift, Archimedean screw, Persian wheels and wind mills are important water lifting devices. A few of the important water lifting devices as practiced by the farmers in villages are discussed here.

10.2.5.1. Swing Basket (Sena)

It is the most ancient water lifting device used in may parts in the country. It consists of a basket or scoop to which four ropes are attached (Fig. 10.9). It is operated by two persons standing face to face with each person holding two ropes in his two hands in same side of the swing basket.

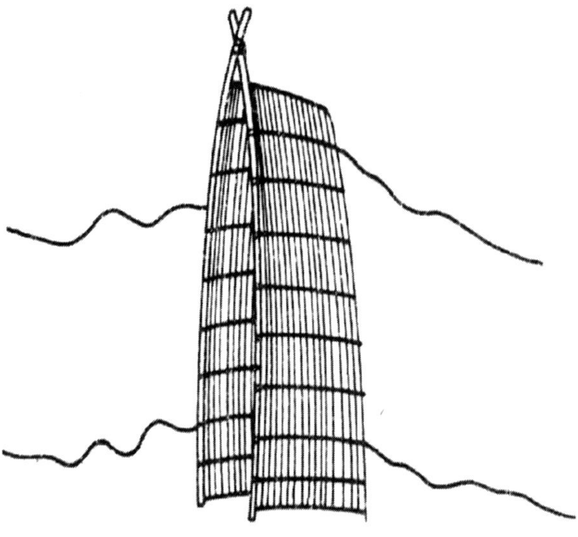

Fig. 10.9. View of the swing basket

They lower the scoops into the water ditch or farm pond and when the scoop is filled with water they lift the scoop to a height where the water has to be discharged to the field or to a channel to lead water to the fields. This is done by manipulation of their hands up and down and to and fro. Water by swing basket can be lifted to a height of 0.8 to 1.5 m. Capacity of swing basket is 8 to 12 lit/swing.

10.2.5.2. Sedua

It is a small device made of a piece of wood which is grooved inside. It is used to lift water to a small height (15 – 25 cm) and the capacity is also very less. Only one person can operate the equipment. It is mostly used to irrigate the vegetables and the discharge capacity is about 1000 lit/hr.

10.2.5.3. Doon/Janta/Karim

It is a boat shaped trough closed at one end and open at the other. The closed end is tied with a rope to a pole to work as a pivot (Fig. 10.10). The open end is hinged to the discharge point. A man stands on a platform above the water source and by applying force he pushes the close end of the doon into the water source. When the doon is filled with water, he lifts the doon by applying jerk by hand so that the water from the doon is discharged to the field through the open end. It lifts water to from a depth of 0.6 to 0.9 m and the discharge capacity varies from 30 to 40 lit. It can irrigate 6000 to 12000 lit/hr.

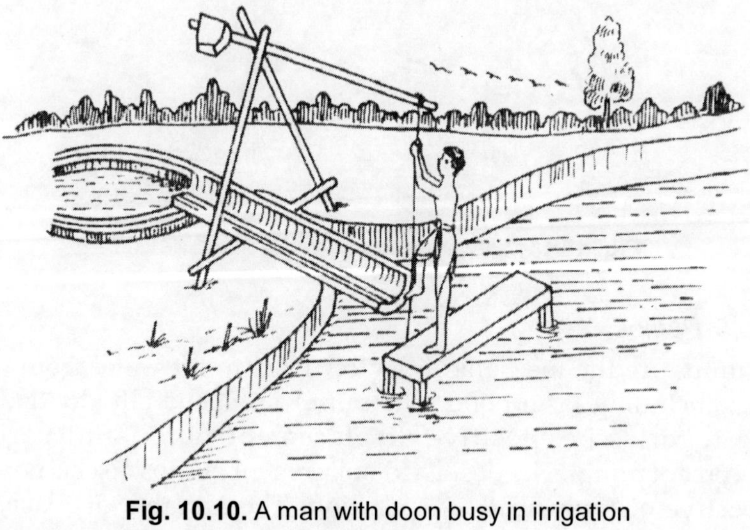

Fig. 10.10. A man with doon busy in irrigation

10.2.5.4. Tenda/Pikota/Dhonkli/Sadoop

It is one of the oldest water lifting device used in ancient countries of the world. It is still used in many villages in India. It is also used in some other countries like China, Japan and Egypt. It consists of a vertical post to act as a fulcrum from which a beam swings up and down like a see-saw. A thin bamboo pole is attached to one end of the beam. A bucket is tied to the other end of the pole. A man at the fulcrum operates the bucket up and down attached to the beam, raises the water full with bucket and then discharges the water into the conveying channels. After the bucket is emptied, it is brought down by giving a jerk with hand and then filled with water by dipping the bucket. Then it is lifted up and water is emptied. The cycle thus continues (Fig. 10.11). Tenda can lift water from a depth 3 to 6 m; the discrage capacity varies from 15 to 25 lit. It can lift water 6000 to 10000 lit per hour from a depth of 6 m.

Fig. 10.11. View of tenda as water lifting device

10.2.5.5. Pumps

Pumps are the mechanical devices used to lift water from water sources including farm pond. The pumps are generally divided into two categories i.e. positive displacement pump and variable displacement pump. The positive displacement pumps discharge water irrespective of their head of operation. The variable displacement pumps have variable discharges rates depending on their heads or depth from which they lift water. The discharge varies inversely with the pressure head. Centrifugal pumps are the most important variable displacement pumps used in agricultural operations.

Centrifugal pumps consists of two basic parts. They are (i) circular impeller and (ii) tight fitting casings. The impeller wheel has 3 to 12 radiating curved vanes. The impeller is set inside the casing. The casing has an inlet pipe which acts as suction pipe and an outlet pipe which acts as delivery pipe. The impeller rotates inside the casing with the help of motor assembly and draws water into the casing. The casing then, by virtue of centrifugal force, throws water from the casing to deliver end. Rotation of the vanes produces high velocity which is then transformed into pressure head. Centrifugal pumps need priming before it starts operating. For priming a foot valve is required. The foot valve is attached to the end of the suction pipe which deeps in water. Foot valve has the following functions:

(i) It helps in building up pressure to start the pump.

(ii) It helps in filtering out the foreign materials which would have otherwise passed through the suction pipe and clogged the pump casing and vanes.

(iii) It also holds water in the suction pipe.

Before priming is done, the suction pipe is fully filled with water and all the air bubles are expelled out. The foot valve should be dipped in depth up to at least 60 cm of water in the source.A centrifugal pump works better when the depth of water in a source is up to 7 m below the pump line and any depth above 8 m, decreases the discharge. The discharge of the pump is inversely proportional to the head. The discharge capacity of the pump varies from 40,000 to 64,000 lit/hr. Details of description, functions and operations of various types of pumps including the centrifugal pumps are available in Michael (1987).

Low Lift Hand Pump

This is a cheap pump and used by many villagers for lifting water from a small depth. It can be operated by hand by a single person. It has a wooden handle connected at one end of the pump bowel and a suction pipe at the other end. The suction pipe dips in the water body and draws water. The delivery pipe is short and connected to the other end of the suction pipe very near to the handle (Fig. 10.12). The delivery pipe cannot throw water to a great distance but delivers water to the conveying channel. When the handle is lifted up, suction is created in the pump chamber/bowel and some water rushes inside the chamber. Then when the handle is lowered down, the water from the pump chamber discharges out. The pump can operate for a low head up to 3 m and has discharge capacity 4000 to 5000 lit/hr. It is ideal to irrigate the crops with low water requirement such as vegetables, flowers, horticultural crops and in kitchen garden. Its weight is not very high being limited up to 40 kg. Therefore it is portable and so can be easily handled by a single person.

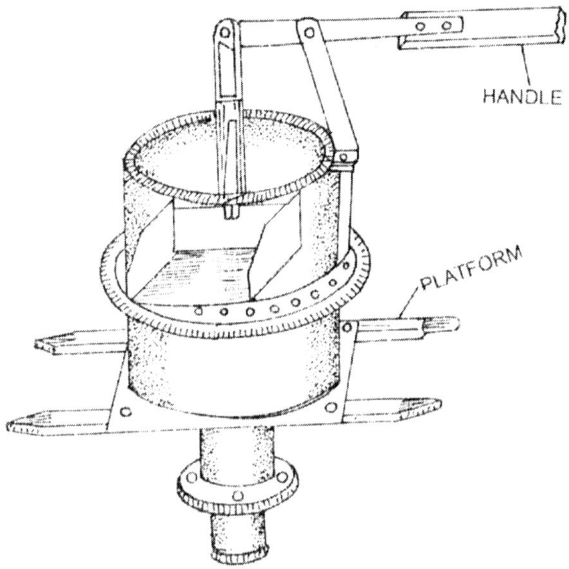

Fig. 10.12. A schematic view of low lift hand pump

Power Requirement of the Pump Motor:

The power requirement of the pump motor is expressed by the term horse power (H.P.) The power of the pump motor relates head with discharge and the expression is given as:

$$\text{H.P.} = \frac{\omega.\ Q.H}{75\eta} \qquad \ldots (10.12)$$

where, w = unit weight of water; Q = discharge to be delivered, cumecs; H = total head against which the pump has to work, m and n = efficiency of the pump set generally taken as 65 percent i.e. 0.65.

The total head is the sum of total suction and delivery head, velocity head and friction losses in pipe including losses at the bends and strainer openings. Velocity head is expressed as:

$$\text{Velocity head} = \frac{V^2}{2g} \quad \text{... (10.13)}$$

where, velocity head is given in metre; V = velocity of water flow in the pipe, m/sec and g = acceleration due to gravity, 9.81 m/sq. sec.

Friction loss is expressed as:

$$h_f = \frac{f'.l.V^2}{2gd} \quad \text{... (10.14)}$$

where, h_f is the frictional loss head, m; l = total length of pipe including vertical and horizontal portions, m; V = velocity of water flow in pipe, m/sec; g = acceleration due to gravity, 981 m/sq. sec and d = diameter of pipe in which water flows, m and f' = coefficient of friction which is generally taken as 0.024.

The losses at the bends and that at the strainer opening is taken as 25 percent of the frictional losses in the pipes.

Sample Calculation

A pump has to irrigate 2 ha wheat crops and the depth of irrigation is 5 cm. The pump runs for 10 hours to irrigate the crop. If the total head against which the pump has to run is 8 m, determine the power requirement of the pump. Assume the pump has efficiency 65 percent.

Solution

Total quantity of water to be supplied for irrigation =2x5=10 ha cm which is equivalent to 1000 m^3.

Time for which the pump runs = 10 hr

Hence, discharge of the pump = 1000 /10 x 3600 = 0.0278 cumec.

Total head = 8 m and the pump efficiency = 0.65

Hence the pump horse power = (0.0278 x 8 x 1000)/ (75 x 0.65) = 4.56 H.P. = 5 H.P.

Hence a 5 H.P. pump can be selected for irrigation.

10.3 Irrigation Management Practice under Deficit Supply

In rainfed areas, availability of irrigation water at right time with required amount to meet the crop water demand is not assured. In this context, one has to judiciously use the limited supply to the crop at the correct time by appropriate method of supply of irrigation so that maximum possible yield and water use efficiency (WUE) are obtained. Values of WUE under field conditions increases, especially in grain crops, with the decrease in soil water supply until it is above a minimum critical level because plants actively try to economies water loss in the range from the minimum critical to the optimum moisture level (Hukkeri and Pandey, 1977). However, total production per unit area decreases as the available soil water comes down below optimum level. Shmuely (1973) has also suggested the application of less than the desired amount of water to growing crops to increase water use efficiency. In areas where water supply is inadequate, limited supplemental irrigation is often practiced with a view to obtain less than maximum yield per unit of available water. The major advantage of the limited supplemental irrigation is the increase of total farm yield and WUE by reducing the area allocated for complete rainfed farming (Stegmen et al., 1980).

Stewart et al. (1976) designed and tested a limited irrigation dryland (LTD) farming system utilizing seasonal rainfall and limited amount of irrigation water. The unique system has flexible adjustment of the amount of land irrigated allowing more land to be irrigated during wet years with a limited amount of irrigation water than during a dry year and it use furrow irrigation without causing runoff from the field.

Correct timing for supplemental irrigation requires knowledge of crop response to applied water at different growth stages. During some particular stages, the plants are more sensitive to moisture stress than at others and this moisture sensitive (critical) periods differ with crops and varieties. In some cases these critical periods may also differ under different climatic and soil conditions for the same crops. These critical moisture-sensitive periods, sometimes called as critical growth stage for various crops has been described in detail by Hukkeri and Pandey (1977) and by Hiler and Howell (1983). Application of water stored in pond as supplemental irrigation at these critical growth stages is more productive than when applied at other stages because moisture stress during these periods will considerably reduce crop yield which can not be recovered by subsequent application of water of later stages.

10.4. Soil Water and Crop Yield Response to Soil Water

There are mainly two crop water production functions models available for assessment of soil water deficit affect on yield of dry land crops at different growth stages. They are (i) multiplicative model and (ii) additive model. Again each of the above two mentioned models may be seasonal or dated. Out of the various types, additive dated model is important. An additive model of dated crop-water production function derived from sensitivity factors for water stress in physiological growth stages as proposed by (Rao *et al.*, 1990) is given as:

$$\frac{Y_a}{Y_p} = 1 - \sum_{tg=1}^{NP} K_{tg} \left(1 - \frac{ETcrop}{PET}\right)_{tg} \quad \text{... (10.15)}$$

where, Ya and Yp = actual and potential yield of the crops, respectively, kg ha^{-1}; k_{tg} = yield response factor for growth stage, tg; ETcrop and PET = actual and potential evapotranspiration, respectively, mm; tg = growth stage index and NP = total number of periods for the crop (generally for all crops it is four). Values of k_{tg} for different crops for different growth stages are reported by Doorenbos and Kassam (1979). Potential yield of the crops is the maximum yield harvested under a set of ideal condition and to be obtained for a particular region from the experimental field observations.

Dated water production function of rice can be used for predicting the yield, but the yield response factors are not available for various stages of its growth. However, the physiological based rice growth and yield model (RICEMOD) developed by McMennamy and O' Toole (1983) can be used to predict the rice yield under practical field conditions.

10.5. Increasing Water Productivity - Some Field Investigations

Numerous experimental data are available on application of supplemental irrigation to different crops at different critical growth stages and their consequential affect on increase in yield, water-use efficiency and economic returns. In this section, the results of a few research works are reported. The result of an experiment at ICRISAT, Hyderabad show that 5 cm supplemental irrigation in alfisol applied to sorghum (at grain-filling stage) and maize (at pre-tasseling stage) increased yields of both the crops by more than 100 percent. However, pear millet showed an increase of 17-59 percent and sunflower 27-32 percent (Anonymous, 1976).

Wani et al. (2003) conducted field experiments on grain-yield response of intercropping and sequential cropping systems to varying doses of supplemental irrigation at ICRISAT, Hyderabad. Good yield response to supplemental irrigation was obtained on alfisols in both rainy and post rainy seasons. The average irrigation water productivity (otherwise called as water-use efficiency or water application efficiency) was observed to be maximum for sorghum (14.9 kg/ha/mm) and for pearl milet, it ranged from 8.8 to 10.2 kg/ha/mm. Tomato gave highest water application efficiency (WAE) with a value of 186.3 kg/ha/mm (Table 10.7). In sorghum + pigeon pea, intercropping system, two supplemental irrigations (SI), each of 40 mm gave additional gross returns of Rs. 3950/ha. The largest additional gross returns from SI was obtained from tomato with Rs. 13,870/ha.

Table 10.7. Grain-yield response (t/ha) of intercropping and sequential cropping systems to supplemental irrigation (SI) at ICRISAT.

Due to one SI of 40 mm			Due to two SI of each 40 mm			
Grain-yield	Increased yield due to SI	WAE* (kg/ha/mm)	Grain-yield	Increased yield due to SI	WAE (kg/ha/mm)	Combined WAE (kg/ha/mm)
			Inter cropping			
	Pearl milet			Pigeon pea		
2.353	0.403	10.0	1.197	0.423	5.3	6.8
	Sorghum			Pigeon pea		
3.155	0.595	14.9	1.22	0.535	6.7	9.4
			Sequential cropping			
	Pearl milet			Cow pea		
2.577	0.407	10.2	0.735	0.425	5.3	6.9
	Pearl milet			Tomato		
2.215	0.35	8.8	26.25	14.9	186.3	127.1

N.B. $\text{WAE} = \text{Water application efficiency} = \dfrac{\text{Increase in yield due to supplemental irrigation}}{\text{Depth of supplemental irrigation}}$

The results of multi-location trials carried out at different locations of India on one supplemental irrigation to rainfed crops are presented in Table 10.8 (Singh, 1983). It is seen from the table that in the northen part of India, wheat has maximum response to supplemental irrigation which varies from place to place. A 5 cm irrigation given at critical stages of wheat, barley, sorghum and rice at 10 different stations of the country increased yield from 15.3 to 23.8 q/ha with efficiency of water use 20.9 kg/ha/mm (Venkateswaralu and Singh, 1982).

Table 10.8. Effect of one supplemental irrigation (5 cm) on yield of rainfed crops.

Crop	Station	No of trials	Yield, q/ha		WUE kg/ha/cm
			Without irrigation	With irrigation	
Wheat	Dehradun	4	21.4	35.5	282
Wheat	Jodhpur	4	19.	36.0	324
Wheat	Ludhina	4	19.2	41.1	431
Wheat	Agra	2	21.9	27.4	110
Wheat	Rawa	4	5.7	18.8	262
Barley	Varanasi	2	26.0	33.6	148
Barley	Ranchi	-	5.5	11.7	124
Sorghum	Bijapur	5	16.5	23.6	148
Sorghum	Bellary	4	4.3	13.7	182
Sorghum	Solapur	5	9.8	18.2	168
Sorghum	Hyderabad	2	13.9	19.2	106
Sorghum	Kovilpatti	3	27.8	28.9	22
Upland rice	Rawa	4	16.2	27.8	232
Bidi tobacco	Anand	4	12.1	18.1	120
Finger millet	Bangalore	-	16.2	23.2	140
Pearlmillet	Jodhpur	4	16.6	19.0	48
Pigeon pes	Jhansi	-	0.5	3.3	56
Sunflower	Bellary	-	1.3	2.9	32
Rapseed	Randhi	-	2.5	5.4	58
Chilles	Bangalore	-	2.1	4.5	48
Bidi tobacco	Vasad	2	14.6	17.5	58
Safflower	Vasad	1	15.6	20.2	92
Mstard	Vasad	1	16.5	20.4	74
Gram	Vasad	1	10.	16.3	110
Cowpea	Vasad	1	20.4	20.8	7.6

Verma *et al.* (1984) have reviewed the available literatures on supplemental irrigation to wheat and maize. Majority of research workers has suggested 5 to 6 cm supplemental irrigation to wheat at early stress especially at pre sowing stage (in case of deficient seeding zone moisture) or at CSRI stage of wheat. In case of maize, tesseling – silking stage has been found most critical for a supplemental irrigation. In northern India an irrigation of 5 cm at pre-sowing or at CRI stage increased rainfed wheat grain yield by 7.7 g/ha (Verma, 1981). According to Gajri et al. (1983) a supplemental irrigation to rainfed wheat increased grain yield by 30 to 81 percent. Time of application, quantity of water for irrigation, and selection of crop types are extensively important factors which need to be considered to derive the maximum benefits. Life saving irrigation at the most crop responsive stage i.e. critical growth stage shall not only protect the crops but help to stabilize the production. The results of supplemental irrigation in medium deep black soils at Bijapur over a period of 5 years indicates that crop yields with one life saving irrigation of 5 cm can enhance the yield by 40-90 percent (Table 10.9, Belgaumi, *et al.*, 1997).

Table 10.9. Response of crops to life saving irrigation in medium deep black soil at Bijapur

Crop	Yield, kg/ha		Yield increase
	Control	Irrigated	(% over control)
Sorghum (Rabi)	1845	2574	39.5
Safflower	1196	1702	42.4
Bengalgram	669	887	32.5
Sunflower	873	1429	63.7
Redgram	198	381	92.4
Hybrid cotton	1094	1524	39.3

Pond irrigation system also plays crucial role in increasing the productivity and income generation capacity of farmers in farming system approach. This system gave higher returns as compared to sole crop of paddy served by pond irrigation. The data obtained at Bijapur on different farming systems revealed that horticultural crops with ber plantations (cv. Umran and Kadaka) recorded gross income ranging between Rs. 27,962 to Rs. 37,260 per hectare with two to three supplemental irrigations as against Rs. 23,657 to Rs. 29,505 per hectare in control from the ber plantation. The next highest income was obtained from the cultivation of curry leaf (Rs. 26, 664 per hectare) with two supplemental irrigations followed by hybrid cotton

(Rs. 18,840 per hectare). Intercropping of bengalgram + safflower in row proportion of 4:2 provided only Rs. 4937 per hectare with two supplemental irrigations where as the income from the arable cropping system was Rs. 3488 under control. Cultivation of mulberry, pastures (grass + legumes) under protective irrigations accounted Rs. 16,870 and Rs. 5107 per hectare, respectively against corresponding income of Rs. 14,385 and Rs. 4121 per hectare under control (Table 10.10, Anonymous, 1991).

Table 10.10. Yield and gross income of different farming systems as influenced by pond irrigation

Farming system		Average yield (kg/ha)	Gross income (Rs./ha)
Ber (Umran)	Irrigated	7452	37260
	Control	5901	29505
Ber (Kadaka)	Irrigated	5592	27962
	Control	4731	23657
Curry leaf	Irrigated	8888	26664
Mulberry	Irrigated	7216	16870
	Control	6245	14385
Pastures	Irrigated	29915	5107
Hybrid cotton	Irrigated	1884	18840
	Control	1501	15010
Bengalgram+Safflower intercropping (4:2)	Irrigated	17260	4937

From the foregoing presentations, it is evident that the response for extensive irrigation with limited supply is more compared to the intensive irrigation. Besides the extensive irrigation, another important consideration to maximize the economic returns from the harvested water of the pond is the method of irrigation. It is advisable to use micro-irrigation techniques for irrigating the crops from the pond stored water. The micro-irrigation like drip and sprinkler has very high application efficiency to the tune of 85-90%. At Bijapur, it has been observed that drip irrigation could save 50% of pond water compared to the surface method of irrigation (Belgaumi, *et al.*, 1997). The following table (Table 10.11) gives the response of a given amount of supplemental irrigation in different fruit crops by different methods of irrigation (Belgaumi, *et al.*, 1997). From this table it is evident that the drip method of irrigation gives highest fruit yield than other methods.

Table 10.11. Response of methods of irrigation on fruit yield of different plants for a irrigation supply of 200lit/plant

Methods of irrigation	Fruit yield, kg/plant			Percent increase over control		
	Ber	Fig	Pomegranate	Ber	Fig	Pomegranate
Drip	6.17	2.73	3.36	252	228	133
Jet	5.07	1.90	2.20	189	128	52
Surface	4.37	1.51	2.31	149	82	60
Control	1.75	0.83	1.44	–	–	–

The selection of catchment and farm pond and the location of the pond in the catchment is very important to maximize the returns from the harvested water. Pond is better to be selected such that there is some catchment area in the upper end of the pond and there is some service/command area below the pond (downstream end of the pond). The crops in the catchment areas may be taken as non paddy crop in both kharif and rabi and the crops in the downstream end of the pond may be paddy in kharif and a non paddy crop in rabi. Selection of the crops and the cropping systems must be location specific. The surface runoff from the upper end of the catchment will be easily harvested in the pond and may not require any supplemental irrigation during kharif and one or two irrigation in rabi. However, the harvested water in the pond can be used as supplemental irrigation to crops grown in the areas at the downstream end of the pond. The irrigation frequency and amount to the crops then will depend on the then available supply and the appropriate water management practice followed for the specific crop. However, proper simulation study need to be undertaken foe selecting the catchment-command area ratio and for deciding the size of the pond. Fig 10.13 gives a representative view of the farm pond constructed in the farm having some catchment area grown with maize and command area grown with rice and other crops in the experimental farm of the department of Agricultural and Food Engineering at IIT, Kharagpur.

To maximize the economic returns from the harvested water of the pond, integrated farm system is advisable. The integrated farming system will help to the increase the incomes from different sources like increased yield of crops from the farms, additional income in rearing fish in the pond, returns from yield of banana, coconut, papaya, arhar etc. grown in the embankment of the ponds. A photo of a pond with cultivation of banana and arhar in the embankment of a newly constructed pond is presented in Fig. 10.14.

Fig. 10.13. View of farm pond having maize in catchment area and rice in command area

Fig.10.15 shows the photo of another farm pond constructed in chronically waterlogged ecosystem in the farm of regional research and technology transfer station of Chiplima, Orissa where fish are reared in the pond and banana and guava crops are grown in the embankment of the pond. In addition, the stored water of the pond is used to irrigate the rice crops. Thus, the pond acts as a viable source for integrated farming system and helps the farmers to raise their income from multi-source.

10.6. Land Levelling Operation for Efficient Use of Water

Before irrigation is practiced in the field, land grading and levelling are required. Land grading is reshaping of the field surface to a planned grade. It is necessary in making suitable field surface to control the flow of water, to check soil erosion and provide for surface drainage.

Land levelling operation may be grouped into three phases: (i) rough grading (ii) land levelling and (iii) land grading.

Fig. 10.14. Banana and arhar grown in embankment of a farm pond

Fig. 10.15. A farm pond for multiuse at Chiplima, Orissa

10.6.1. Rough Grading

It constitutes the removal of abrupt irregularities such as mounds, dunes, and ridges and filling of pits, depressions and gullies.

10.6.2. Land Levelling

It is well known that, the agricultural lands have non-uniform slopes and many undulations, which make the condition of soil surface difficult to supply the irrigation water uniformly and efficiently for irrigating the crops. In this situation, the necessity of land leveling or grading becomes very essential. In land grading/levelling operations, the land surfaces are prepared or modified to a plane grade, to provide a suitable top soil surface for efficient application of irrigation water and also to facilitate quick disposal of excess water from the area. The proper application of irrigation water refers to repellents the moisture content up to the root zone depth. The water applied more than the required, is not only a loss but also results the development of water logging condition, which damages the crop, and virtually reduces the crop yield. In this way, none of the irrigation method except micro-irrigation is able to supply the irrigation water uniformly all over the field. The land leveling/grading or land shaping operations remove this kind of irregularities up to a large extent.

10.6.3. Land Grading

The operation in which land is converted to a desired grade in longitudinal as well as across-slope directions for efficient water application is called land grading.

10.6.3.1. Objectives of Land Grading

The main objectives of land grading are as under:

1. To apply irrigation water very uniformly throughout the field.
2. To economize the utilization of water by minimizing the losses.
3. To augment the surface drainage.
4. To apply the fertilizers, uniformly throughout the field.
5. To increase the crop yield by uniform application of water.
6. To have uniform plant population
7. To save the labour and water

10.6.3.2. Factors Affecting Land Grading

The land grading operations are performed in the field on segment basis. Therefore, before starting the grading work, it is very essential to prepare the farm layout in respect of location of field channels, drains farm roads etc. The main factor affecting land grading is soil profile limitation. The nature of top soil, depth of sand, gravel or hard pan layer, water table depth etc. are taken into consideration under soil profile limitations, for planning of land grading work. The depth of gravel or hard pan affects the depth of cut,. However, the depth of cut should be such that it can not expose the inert materials which are not suitable for good crop. For good cultivation, it is essential to have at least 1.0 m soil depth. Hardpan or sandy layer should not be exposed. In large areas, the land may be divided into several strips across the main slope and then individual strips should be handled for grading work. The objective is achieved by removing the topsoil and placing at some other places and when land grading work has been finished, than collected topsoil should be uniformity spread throughout the area. The soil depth should always be in such extent in which roots of the crops could be easily accommodated. If the sub soil or sub-stratum is coarse textured (sand or gravel), which have high infiltration rate and hydraulic conductivity, a higher grade is recommended so that the water could run rapidly to the lower reaches of field.

The land slope and soil texture are other soil profile factors that affect the land grading. The steep slope with undulating terrain and shallow soil depth make difficult to conduct the lelvelling work. A soil composed of coarser particles has greater infiltration rate resulting into rapid absorption of water. In this condition, a greater land slope is recommended. To the contrary, fine textured soils have lesser infiltration rate; as a result smaller grade is provided in such soils.

The land levelling and grading are the expensive operations due to involvement of heavy earthworks. However, the cost could be minimized by reducing the volume of earthwork and haulage distance by proper planning of land levelling/grading operations technique. The very purpose of land levelling, grading and smoothing is to help the irrigation water to be uniformly distributed over the whole area with less application losses. Increase of application efficiency depends on management factors including provision of proper irrigation and cross slope, deciding optimum plot size, soil and crop management factors and adopting suitable irrigation scheduling. In land grading work, two types of slopes are prepared, in which one is irrigation slope and other is cross slope.

10.6.3.3. Irrigation Slope

It is the slope given to the field for running the irrigation water in forward direction. Sometimes, this slope is also referred as longitudinal slope. Its range depends on the soil type, stream flow rate etc. In coarse textured soil, greater irrigation slope is recommended than the fine textured soils because greater slope minimizes the time of absorbing of water by the soil,. In addition, the irrigation slopes should also be in such a limit that the running water in the border or furrow could not cause soil erosion problem. However, based on soil types, the recommended safe limits of longitudinal slope are given in Table 10.12.

Table 10.12 Recommended longitudinal slopes based on soil type

S. No	Soil type	Longitudinal slope, %
1	Heavy soil (clay)	0.05-0.25
2	Medium soil (Loam)	0.20-0.40
3	Light soils(Sandy)	0.25-0.65

10.6.3.4. Cross Slope

Cross slope is perpendicular to the longitudinal slope. The limit of permissible cross slope depends on the method of irrigation used. For example, the borders and checks do not allow having the cross slope within the boundary while furrows and corrugations require a certain degree of cross slope. Generally, the cross-slope is being equal to or little less than the longitudinal slope. The maximum slope should be up to 0.5% or less in alluvial soils. Generally for eliminating the cross-slope an excessive soil cut is required, which could be reduced by dividing the entire field into different compartments and each compartment is levelled at different elevations, separated by low ridges. The volume of earthwork evolved in land grading work depends on the magnitude of diagonal slope at right angles to the direction of irrigation.

In land grading works for surface drainage, the longitudinal slope can be varied, as per requirement to control the soil erosion, reduce yardage or to avoid the exposing of extensive area of the suib soil. Planned grading should be such that each row would carry its share of surface run-off from upstream to downstream part without overtopping into the next row. The graded surface should be sloped to a specified design grade. The grade in the direction of drainage may be uniform or variable with a reverse grade. A minimum grade of 0.1% should preferably be used, if possible. However, the limit of maximum grade is decided on the basis of erosion hazard.

The magnitude of irrigation and cross slopes is decided on the basis of method of irrigation to be used. The characteristic features for the basin irrigation are:

- Soil: Medium fine textured and moderately coarse textured soil.
- Crops: Vegetable crops and paddy.
- Irrigation slope: 0.6%
- Cross slope: 0.2%

The characteristic features for the boarder irrigation are:

- Soil: Medium to fine textured soils
- Crops: Cereal and row crops.
- Irrigation slope: 0.6%
- Cross slope: 0.2%

The range of land slopes, boarder length and stream flow rate vary from soil to soil depending on the texture of soil (Table 10.13).

Table 10.13. Recommended border length, stream flow rate and slopes for different soil

Soil Texture	Border length, m	Stream flow rate, m^3/h	Land slope, %
Coarse texture	60-90	20-25	0.40-0.6
Moderately coarse texture	76-150	15-20	0.4-0.6
Medium texture	90-180	13-18	0.4-0.6
Moderately fine texture	90-180	10-14	0.4-0.6
Fine texture	100-200	6-8	0.2-0.4

10.7 Performance of Farm Irrigation Systems

The performance of farm irrigation systems is determined by the efficiency with which water is delivered to the field and utilized to the fullest amount possible with minimum losses due to seepage and percolation, runoff, evaporation and other operational losses. The efficiency of the irrigation system will increase if the conveyance losses, application losses and other losses are reduced to a great extent.

The overall efficiency of a farm irrigation system is defined as the percent of water applied to the farm that is used for irrigation on the farm. Irrigation efficiency otherwise called as overall system efficiency is defined (James, 1988) as:

$$E_i = \left(\frac{S - DP - SR - O}{S}\right)100 \qquad \text{... (10.16)}$$

where, E_i = irrigation efficiency, percent; S = amount of water supplied to the farm; DP = deep percolation loss; SR = surface runoff of the farm and O = operational losses due to conveyance and storage system.

Sample Calculation

A measuring structure is installed at the inlet of a canal system. The structure measures 1.5 cumecs of water and water is derived to the canal system for 10 hours continuously. If during 10 hours of operation, deep percolation loss = 2000 m^3; surface runoff = 3500 m^3 and operational losses = 1000 m^3 then determine the irrigation efficiency of the system.

Solution

Amount of water delivered to the farm during 10 hours = 1.5 x 10 x 3600 = 54000 m^3.

Operational losses = 1000 m^3

Surface runoff = 3500 m^3

Deep percolation loss = 2000 m^3

$$\text{Hence, irrigation efficiency} = \left(\frac{54000 - 1000 - 3500 - 2000}{54000}\right)100$$

$$= 87.96 \text{ percent.}$$

Thus, 87.96 percent of water applied to the farm is used by the crops and rest is lost due to surface runoff, deep percolation loss and operational losses.

When evaluating the performance of a farm irrigation system, it is useful to examine the efficiency of each system components. The main system efficiency components are conveyance efficiency, storage efficiency and application efficiency. The overall system efficiency is the product of above mentioned three system efficiencies and is given as:

$$E_i = \left(\frac{E_s}{100}\right) \times \left(\frac{E_c}{100}\right) \times \left(\frac{E_a}{100}\right) \times 100 \qquad \text{... (10.17)}$$

where, E_i = irrigation efficiency; E_s = water storage efficiency; E_c = conveyance efficiency and E_a = application efficiency. All the efficiency terms are expressed in percent.

10.7.1. Water Storage Efficiency

The water storage efficiency in a farm pond gets reduced due to the losses like evaporation and seepage. The following formula can be used to evaluate the storage efficiency of a farm pond.

$$E_s = \left(1 - \frac{V_s + V_e}{V_i}\right) 100 \qquad \text{... (10.18)}$$

where, E_s = storage efficiency; V_i = inflow volume entering a farm pond; V_s = seepage volume from the pond and V_e = evaporation volume of the pond.

10.7.2. Water Conveyance Efficiency

Total amount of water that is delivered from a farm pond never reaches a field since on the way there are losses in the channels. These losses are called as conveyance loss. The conveyance losses occur due to the earthen channels or ill maintained channels. However, by lining the channels, the conveyance losses can be reduced and so the conveyance efficiency can be increased. The conveyance efficiency is expressed as:

$$E_c = \frac{V_{df}}{V_{dp}} \times 100 \qquad \text{... (10.19)}$$

where, E_c = conveyance efficiency; V_{df} = volume of water delivered to the field and V_{dp} = volume of water diverted from the pond.

10.7.3. Water Application Efficiency

Water application efficiency for an irrigated area is defined as the ratio of the volume of water actually used by the plants to the volume of water delivered to the field. It is to be noted that the amount of water supplied to each filed is not entirely used by the crops. Some amount of the supplied water is lost due to seepage and percolation; some is lost due to runoff and other unavoidable losses. By economic application of irrigation water and scientific water management, the application efficiency can be increased. Water application efficiency is expressed as:

$$E_a = \frac{V_{ac}}{V_{df}} \times 100 \qquad \text{... (10.20)}$$

where, E_a = water application efficiency, V_{ac} = volume of water actually used by the crops and V_{df} = volume of water delivered to the field.

10.7.4. Water Distribution Efficiency

Some times another efficiency term is used in irrigation system. That is called as water distribution efficiency and is define as:

$$E_{de} = \left(1 - \frac{\bar{y}}{\bar{d}}\right) 100 \qquad \text{... (10.21)}$$

where, E_{de} = water distribution efficiency; $\bar{d}$ = average depth of water stored along the run during irrigation and $\bar{y}$ = average numerical deviation from $\bar{d}$.

10.7.5. Water Use Efficiency

In addition to the above mentioned efficiencies, often another efficiency called as water use efficiency (sometimes also called as field water use efficiency) is used. Water use efficiency is expressed as:

$$WUE = \frac{Y}{WR} \times 100 \qquad \text{... (10.22)}$$

where, WUE = water use efficiency; Y = yield of the crop and WR = water requirement of the crop. Instead of WR if actual evapotranspiration term (ETcrop) is used in the denominator of water use efficiency, then it is designated as crop water use efficiency (CWUE).

Sample Calculation

On a certain day, 1000 m^3 of water is diverted from a farm pond. Out of this, 200 m^3 is lost as seepage loss in the conveyance system. Find out the conveyance efficiency of the system.

Solution

Given the volume of water diverted from a farm pond = 1000 m^3.

Conveyance loss = 200 m^3

So, water actually reaching the field = 1000 200 = 800 m^3

Therefore, conveyance efficiency = 800 / 1000 = 0.8 = 80 percent.

Sample Calculation

A stream of 135 lit/sec was diverted from a canal and 100 lit/sec were delivered to the field. An area of 1.6 ha was irrigated in eight hours. The effective root zone depth of the crop was 1.8 m. Runoff loss of the field was 432 m^3. Depth of water penetrated in the soil linearly varied from 1.8 m at up end to 1.2 m at down end of the field. Available moisture holding capacity of the soil is 20 cm/m depth of the soil. Determine the water conveyance efficiency, water storage efficiency, water application efficiency and water distribution efficiency.

Solution

Amount of water diverted from the canal = 135 lit/sec

Amount of water delivered to the field = 100 lit/sec

So, the conveyance efficiency = 100/135 = 0.74 = 74 percent.

Water delivered to the field = (100 x 60 x 60 x 8) /1000 = 2880 m^3

Water lost as runoff = 432 m^3

So, water stored in the root zone of the crop = 2880 432 = 2448 m^3

Hence, water application efficiency = 2448/2880 = 0.85 = 85 percent.

Water holding capacity of the root zone = 20 x 1.8 = 36 cm.

Moisture required in the crop root zone = $36 - \left(\frac{36 \times 50}{100}\right)$ = 18 cm

Volumetrically the above amount equals = (18 /100) x 1.6 x 10000 = 2880 m^3

Hence, water storage efficiency = 2448/2880 = 0.85 = 85 percent.

Average depth of water stored along the run during irrigation = (1.8 + 1.2)/2 = 1.5 cm.

Numerical deviation from depth of penetration = {(1.8 -1.5) + (1.5-1.2)}/2 = 0.3 cm.

Water distribution efficiency = (1-0.3/1.5) x 100 = 80 percent.

Question Banks

Q1. Briefly describe the various methods by which stored water in farm pond can be better utilized.

Q2. What do you mean by multiuse of water? Explain how a farm pond can be used as a viable source for multi-use of water.

Q3. Describe the two important options to utilize the stored water in the farm pond for irrigation.

Q4. What is the difference between the intensive and extensive irrigation? Briefly describe the merits and demerits of each system.

Q5. What do you mean by irrigation scheduling? What are the parameters used to define the irrigation scheduling?

Q6. Write the physiological and critical growth stages of crops mentioned below:

(i) Rice (ii) Wheat (iii) Maize (iv) Potato (v) Tomato (vi) Sorghum (vii) Ground nut and (viii) Mustard.

Q7. How is the irrigation timing decided based on the criteria of soil moisture level?

Q8. What is IW:CPE approach in irrigation scheduling? How does the irrigation timing get decided by this approach?

Q9. A crop of chilli is irrigated based on the approach of IW: CPE. Irrigation is given with fixed value of IW: CPE of 1.2. If 6 cm of water is supplied at each irrigation, then find out at what value of CPE, the irrigation is to be applied to the crop? If there is a constant rate of pan evaporation of all the days, then also find out after how many days, irrigation has to be supplied to the crop?

Q10. A crop of potato is grown in Sambalpur and the following data were recorded. Determine the time at which irrigation has to be supplied to the crop based on available soil moisture depletion study.

Field capacity, F C = 22.0% dry weight basis; permanent wilting point, PWP = 10% dry weight basis; bulk density, BD = 1.40 g/cc; effective root zone depth, RD = 60 cm; daily water use, ETcrop = 0.25 cm/day and allowable soil water depletion = 25 percent of available soil water.

Q11. Write the formula to compute the irrigation amount to be given to a crop based on moisture depletion level?

Q12. Find out the net quantity of irrigation water to be applied to a crop of potato with the following values of soil moisture content and other data.

Depth of soil layer in (cm)	Moisture % on oven-dry basis		Apparent specific gravity, g/cc
	Field capacity	Actual	
0 - 15	28.0	19.4	1.39
15 - 30	26.0	17.8	1.47
30 - 45	22.3	19.2	1.51
45-60	22.2	20.5	1.50

Q13. Find out the gross irrigation amount of a crop if the net amount of irrigation is 5 cm and the field efficiency is 50 per cent.

Q14. What are the different methods of irrigation? How does the method of irrigation help in scheduling of irrigation?

Q15. What is a surface method of irrigation? Briefly mention the various merits and demerits of surface irrigation.

Q16. What do you mean by wild flooding or free flooding? What are the advantages and disadvantages of free flooding?

Q17. Describe the methods of irrigation with limitations for important field crops.

Q18. How can the water productivity of a system be enhanced under deficit irrigation supply?

Q19. What is scientific water management practice in crops? How does it help in increasing the water use efficiency of crops?

Q20. What is meant by water productivity? Describe the steps to be taken to increase the water productivity at field levels?

Q21. How does the cropping system and water management practice help in increasing the water productivity at the farm level?

Q22. What is a dated crop water production model? Write the additive model of crop water production and explain how is it used to estimate the actual crop yield?

Q23. Calculate the actual yield of crop to be obtained by using the additive model of dated crop water production function with the help of the following data.

Parameters/Crop growth stage	Initial	Crop development	Mid season	Maturity stage
Actual evapotranspiration, mm	10	50	62	32
Potential evapotranspiration, mm	12	60	72	40
Crop stress factor	0.5	0.55	0.75	0.5

The maximum yield of the crop is observed to be 10 q/ha.

Q24. What is a protective or life saving irrigation? Which is the most important crop growth stage at which protective irrigation is to be supplied?

Q25. What is a water lifting device? Write notes on the following water lifting devices.

i) Tenda (ii) Doon (iii) Swing basket and (iv) Low lift hand pump.

Q26. Write the formula to determine the horse power of a motor driven pump.

Q27. A pump runs for 10 hours to irrigate a crop in 0.3 ha and the irrigation requirement of the crop is 6 cm. The pump has efficiency 65 percent. The diameter of both the suction and delivery pipes is 5 cm. The lengths of the suction and delivery pipes are 3 and 2m, respectively. Calculate the horse power of the pump?

Q28. Define and explain the following efficiencies:

(i) Water conveyance efficiency (ii) Water storage efficiency (iii) Water application efficiency and (iv) Water distribution efficiency.

Q29. How can the overall irrigation system efficiency be improved?

Q30. A stream of 125 lit/sec was diverted from a canal and 100 lit/sec were delivered to the field. An area of 1.5 ha was irrigated in 10 hours. The effective root zone depth of the crop was 1.5 m. Runoff loss of the field was 430 m^3. Depth of water penetrated in the soil linearly varied from 1.6 m at up end to 1.2 m at down end of the field. Available moisture holding capacity of the soil is 20 percent. Determine the water conveyance efficiency, water storage efficiency, water application efficiency and water distribution efficiency. .

Q31. Describe the layout plan of sprinkler and drip irrigation.

Q32. How can the performance of a sprinkler irrigation system be evaluated?

Q33. What is the importance of discharge measurement in the fields? Describe how Parsdhall and cutthroat flumes are used to measure discharges in the channels?

Q34. A Parshall flume is used to measure discharge in a channel for irrigating wheat crop. The throat width of the flume, W = 7.5 cm; head measured in the converging and diverging section of the flume are 40 and 28 cm, respectively. Find out the discharge measured by the flume.

Q35. Find the discharge measured by a cutthroat flume installed in a channel for irrigation to mustard crop with the following recorded data.

Length of flume = 4 ft; width at throat section = 3 ft; upstream head = 2.2 ft; downstream head = 1.8 ft.

Q36. How does land leveling and land grading help in increasing the efficient utilization of irrigation water?

Q37. Describe the various factors that affect the land grading?

Q38. How does the irrigation slope vary with types of soil? Also explain the factors to decide the cross slopes?

Q39. Write the characteristic features of both basin and border irrigation. Mention the crops for which basin and border irrigation methods are employed.

Q40. If the yield of a crop due to five supplemental irrigation of 5 cm each is 2.2 ton/ha, then find out the irrigation water use efficiency? For the same yield, if the total water requirement of the crop including irrigation is 35 cm, then find out the field water use efficiency of the crop.

Q41. For a given crop with same yield, which one of the following two efficiencies will be more and why?

I. Field water use efficiency and (ii) Crop water use efficiency.

Q42. Determine the crop water use efficiency of rice which has a yield of 3.5 ton/ha. The values of crop coefficient and reference evapotranspiration of the crop in different growth stages are as follows.

Parameters/Crop growth stage	**Initial**	**Crop development**	**Mid season stage**	**Maturity**
Reference evapotranspiration, cm	10	16	22	9
Crop coefficient	0.95	1.1	1.15	0.9

References

Anonymous. 1976. Annual Report of International Crop Research Institute for the Semi-Arid Tropics, Hyderabad, pp. 80.

Anonymous. 1991. Annual Progress Report of All India Co-Ordinated Research Project on Dryland Agriculture, Bijapur.

Anonymous. 1997. Status of Research on Agricultural Water Management (Eastern Region), AICRP on Water Management, Chiplima Centre, Orissa University of Agriculture and Technology, Bhubaneswar.

Belgaumi, M.I., Itnal, C.J. and Radder, C.D. 1997. Farm Ponds. Technical series 4, Publication Centrte, University of Agricultural Sciences, Dharwad, pp. 47.

Doorenbos, J. and Kassam, A.H. 1979. Yield Response to Water. Irrigation and Drainage Paper 33, FAO, Rome.

Doorenbos, J. and Pruitt, W.O. 1977. Guidelines for Predicting Crop water Requirements. Irrigation and Drainage Paper 24, FAO, Rome

Gajri, P.R., Verma H.N. and Parihar, S.S. 1983. Rain water harvesting and its recycling for maximization of crop production. Indian J. Soil Conserv., Vol. 9(2&3): 116-123.

Hiler, E.A. and Howell, T.A. 1983. Irrigation options to avoid critical stress. An overview in limitations to efficient water use in crop production. Madison, Wis. Soil Sc. Soc. Am.

Hukkeri, S.B and Pandey, S.L.1977. Water requirement and irrigation of crops. In: Water Requirement and Irrigation Management of Crops in India. Monograph No. 4, New Series Water Technology Centre, IARI, New Delhi: 163-273.

James, L.G. 1988. Principles of Farm Irrigation System Design. John Wiley & Sons, New York, pp. 543.

McMennamy, J.A and O' Toole. J.C. 1983. RICEMOD: A Physiologically Based Rice Growth and Yield Model. International Rice Research Institute, Manila.

Michael, A.M. 1987. Irrigation Theory and Practices. Vikash Publishing House Pvt. Ltd., New Delhi.

Misra, R.D. and Ahmed, M. 1987. Manual on Irrigation Agronomy, Oxford & IBH Publishing Co. Pvt. Ltd., New Delhi, pp. 412.

Rao, N.H., Sarma, P.B.S. and Chander, S. 1990. Optimal multicrop allocation of seasonal and intraseasonal irrigation water. Water Resour. Res., Vol. 26(4): 551-559.

Shmuely, E. 1073. Efficient utilization of water in irrigation in arid zone irrigation. Chapmen and Hall Ltd.:411-423.

Singh, R.P. 1983. Farm Pond. Project Bulletin No. 6, All India Co-ordinated Research Project on Dryland Agriculture, Hyderabad.

Stegman, E.C., Musick J.T. and Stewart, J.I. 1980. Irrigation Water Management. In: M.E. Jensen (Ed.), Design and Operation of Irrigation System, ASAE Monograph, St. Joseph, Michigan: 763-816.

Stegmen , E.C.; Musick, J.T. and Stewart, B.A. 1980. Irrigation water management. In Design and Operation of Farm Irrigation system, Ed. by Jenson, M.E.., ASAE: 761-816.

Stewart, J.I., Hagan, R.M. and Pruitt, W.O. 1976. Production functions and predicted irrigation programmes for principal crops as required for water resources planning and increased water use efficiency. Final Rep., U.S. Department of Interior, Washington DC, 80 pp.

U.S. Bureau of Reclamation 1975. Water Measurement Manual , U.S. Government Printing Office, Washington, D.C. pp. 327.

Venkateswaralu, J ans Singh, S.N. 1982. Crop management to mitigate moisture stress. In a decade of dry land agriculture research in India (1971-80). All India Co-Ordinated Research Project on Dryland Agriculture, Hyderabad: 48-50.

Verma, H.N 1981. Water harvesting for life saving irrigation of rainfed crops in the sub-mountain region of Punjab. J. of Agric. Engrg. (Indian Soc. of Agril. Engineers), Vol. 18(3&4): 64-72.

Verma, H.N., Prihar, S.S. and Singh, R. 1984. Feasibility of storage of runoff in dugout ponds and its use for supplemental irrigation in submontane Punjab. Indian J. Soil Conserv., Vol. 12(1): 31-36.

Walker, W.R and Skogerboe, G.V. 1987. The Theory and Practices of Surface Irrigation. Prentice-Hall, Inc., Englewood Cliffs, N.J.

Wani, S.P; Pathak, P; Sreedevi, T.K; Singh, H.P. and Singh, P. 2003. Efficient management of rainwater for increased crop productivity and ground water recharge in Asia. Comprehensive Assessment of Water Management in Agriculture Series, Water Productivity in Agriculture: Limits and Opportunities for Improvement. Book Edited by J.W.Kijne, R. Barker and D. Molden. CABI Publishing, pp. 352.

Zimmerman, J.D. 1966. Irrigation. John Wiley and Sons, New York, N.Y.: 203-253.

Chapter 11

Economics of Farm Pond Irrigation Systems

11.1 Introduction

Economic analysis is the most important parameter for determining the optimum size of farm pond. It is, no doubt that a big size farm pond can meet the irrigation demand of the crops grown in the command areas in addition to meeting the other demand in the command areas. But a big size farm pond is an oversize structure which necessitates large investment to be incurred which may not be economically viable. On the other hand, a small size structure may not be able to meet the crop water demand and also other demands in the command areas and so is called as under-designed structure. It has been observed that as the size of the farm pond gradually increases from a certain minimum size, the total investment cost increases. At the same time, the benefits from the increasing storage capacity of the farm pond also get increased. In such cases, from a purely business point of view, the most economical capacity is that which provides the highest rate of return on investment. As the size of the farm pond increases, the proportion of the benefit/profit to the cost also increases from certain minimum size, attains maximum value at a particular size and then declines as the size further increases. This particular size of the farm pond at which the net profit is maximum, is called as the optimum size. It is to be noted that at the optimum size, the other economic indicators like the benefit-cost ratio and internal rate of return are also high. The variation of net profit with size of the farm pond is a concave curve. The size of the farm pond corresponding to the peak of the curve which represents the highest net profit is the optimum size of the pond.

The principle of obtaining the optimum size of farm pond is to first assume a certain minimum size say five percent of the farm area and conduct the economic analysis by studying the costs and net profits of the assumed size. Then the size is gradually increased and the economic analysis is conducted for each increased size. Then a curve relating the size of the farm pond and net profit is drawn. Such a curve is generally concave shaped. The size of the structure corresponding to the maximum net profit gives the optimum size. It is to be noted that any size chosen more or less than the optimum size would give less net profit and benefit-cost ratio than that of the optimum size.

Economic analysis has been recommended by a number of researchers to determine the optimum size of the farm pond in rainfed areas (Panigrahi, 2001; Panigrahi et al., 2007; Verma and Sarma, 1990). In any economic analysis, all cash flows must be evaluated at some reference time. A present worth analysis should be used for the study (Taylor, 1975) since all cash flows over the life of the system can thus be converted, using appropriate factors to account for interest and inflation into an equivalent present value. In the economic analysis, following items are to be considered: (a) Initial investment, (b) Pond maintenance cost, (c) Irrigation cost, (d) Land lease cost, (e) Production costs of the crops and (f) Annual returns from irrigation to grown crops in pond service area, returns from crops grown in pond embankment and returns from rearing fich in the pond etc. It should be assumed that all inputs except irrigation are same for both irrigated and rainfed crops.

11.2 Initial Investment

Initial investment of pond irrigation system, also called as fixed cost includes the following components:

- Construction cost of pond
- Lining cost of pond if lining is required and
- Labour cost of lining (for lined pond).

Construction cost of pond is computed by knowing the volume of excavation of earth materials for construction of the pond. For a particular size of the pond with assumed depth and side slope, the dimensions of farm pond are computed as discussed in Chapter 6 "Dimensions of Farm Pond". Then the volume of the pond is computed by using the prismoidal formula (Eqns. 6.7 and/or 6.8) which gives the volume of earth work to be excavated. It is to be remembered that for same volume of earth work excavation, a pond having higher depth would require more cost of excavation than the pond with lower depth. This is because a deeper pond is more difficult to be constructed since lifting of excavated soil is more difficult and need more labour costs than a shallow pond. Similarly the cost of excavation also depends on the type of soil. If the soil of the pond site is soft, then digging becomes easier and the cost of excavation is less. On the other hand if the soil is hard and contains rock, then excavation cost need to be more. The lining cost of the pond depends on the amount of areas to be lined and the type of the lining materials. The prevailing rate of lining materials are to be taken into account while deciding the lining cost. Cost of brick masonry and plastic lining is more than the ordinary soil-cement or clay lining. The total labour cost/wages for lining the ponds again depends on the amount of areas need to be lined and the type of lining. The prevailing schedule rate of labour cost of specific areas should be taken into account while evaluating the economic of developed irrigation system.

11.3 Variable Cost

Variable cost also called as annual operating cost includes the following costs:

- Pond maintenance cost
- Land lease cost to construct the pond
- Irrigation cost to meet the irrigation supply from the pond to the crops and
- Cost of production of crops using irrigation from pond.

Maintenance cost is used to desilt the farm pond and maintain and repair the spillways and embankments of the farm pond every year. Pond maintenance cost depends on the years of service of the pond, the operating environment and the quality of maintenance. In addition, there are substantial variations in the prices paid for parts and supplies, and wages paid to the maintenance personnel. Annual maintenance costs should therefore be based on local data wherever and whenever possible. When the local data are not available then the annual maintenance cost of farm pond irrigation system can be assumed as a constant yearly cost that may be taken as 2% of initial investment of pond (Palmer et al., 1982; Mishra et al., 1998). For different pond sizes, the initial investment will be different and hence, the maintenance cost will also be different.

For construction of pond land may be taken on lease. The existing lease rate of land in the area specific to be constructed with pond should be considered. Irrigation cost depends on the method and amount of irrigation which includes the pumping cost including the energy cost and irrigation conveyance cost by pipes. If drip and sprinkler are used for irrigation then the same must be considered in the irrigation cost. For various pond sizes, application of supplemental irrigation to crops grown in the command areas of the farm pond will be different and hence, irrigation cost will also be different. The cost of production of crops depends on the acreage of the crop and the type of the crops grown. The production cost for different crops are different and also the rate varies from region to region. Hence, the production cost of the areas specific must be considered. Production costs include the costs of (i) land preparation (ii) sowing/planting (iii) intercultural operation (iv) fertilizer and pesticide application (v) harvesting. It also includes the input costs like seeds, fertilizer and pesticides.

11.4 Annual Returns from Irrigation

It is obvious that crop yield show a positive response to the enhanced soil moisture effected due to the supplemental irrigation from the farm pond. Annual returns are the values of the increased yields of the crops due to supplemental irrigation. This is calculated by knowing the minimum support price of the products of different crops grown (both grain/seed and their byproducts) and the increased yield due to irrigation. Total annual returns are to be calculated by adding the returns due to fish cultivation and growing different crops in the farm areas as well as in the embankments of the pond. For various sizes of farm ponds, the amount of water available for supplemental irrigation to crops will be different and hence the yields of crops will be different as discussed in section 10.4. i.e. "Soil water and crop yield response to soil water" in Chapter 10. The increased yields from the increased size of farm pond will fetch higher annual returns which will influence the economic indicators also.

11.5 Present Worth Analysis

Present worth is the amount that must be invested at the beginning of the analysis period to return the equivalent of a component's initial cost plus interest by the end of the analysis period. As proposed by Palmer et al. (1982), present worth analysis is to be used to evaluate all the cash flows in order to account for the interest and inflation factor in investment. Considering the yearly inflation (f) on total yearly cost (A), the cost in year n read, as B would be:

$$B = A(1+f)^{n-1} \quad \text{... (11.1)}$$

where, A = total yearly cost that is also called as the variable cost and is given as the summation of pond maintenance cost, land lease cost and irrigation cost. To project each yearly cost with consideration of inflation rate (B) back to the present from year n, using an interest rate (i), that is termed as present worth (PW), following equation was used.

$$PW = B(1+i)^{-n} \quad \text{...(11.2)}$$

Combining equation (11.1) and (11.2) and summing over n year of pond life system, the present worth of total yearly cost A i.e. PW_A is:

$$PW = \sum_{i=1}^{n} A(1+f)^{n-1} . (1+i)^{-n} \quad \text{... (11.3)}$$

The total cost of the irrigation system ($Total_{cost}$) is:

$$Total_{cost} = I_{Inv} + PW_A \quad \text{.... (11.4)}$$

In the similar way of total yearly cost, present worth value of total annual returns from irrigation (PW_{RI}) is computed and is represented as:

$$PW_{RI} = \sum_{i=1}^{n} R\,(1+f)^{n-1} .(1+i)^{-n} \qquad \text{... (11.5)}$$

where, R = total annual returns from irrigation and other terms have been defined earlier.

In the economic analysis inflation rate or called as annual rate of cost of escalation should be the average over several years' data since it varies from year to year and also varies from country to country. Similarly the farmers may borrow money from the lending agencies for the installation of the system and the interest rate for the agricultural purpose also varies from time to time and country to country. Hence, an average interest rate over the whole analysis period may be used as the rate of interest. Economic analysis of on-farm pond irrigation systems typically ranges with life span of pond varying from 20 to 30 years whereas for large complex projects, it ranges from 40 to 100 years (Thompson *et al.*, 1980).

11.6 Economic Indicators

In any economic analysis generally four main indicators are studied. These indicators are defined as below:

(i) Net Profit

The net profit (NP) due to supplemental irrigation from farm pond is given as:

$$NP = PW_{RI} - Total_{cost} \qquad (11.6)$$

(ii) Benefit-Cost Ratio

Benefit cost ratio (BCR) is another aspect of study of economic analysis. This is calculated (Islam *et al.*, 1998) as:

$$BCR = \frac{PW_{RI}}{Total_{cost}} \qquad (11.7)$$

A project is considered to be economically viable if there is net profit i.e present worth of the returns is more than the total project cost and value of *BCR* is more than 1.0. If the value of *BCR* is 1.0, then there is no loss and no profit and a value less than 1.0 indicates that the project is not justified.

(iii) Pay Back Period

Pay back period *(PBP)* is the period in which the project cost is paid back to the lending agencies. It should be less than the life span of the structure. Generally a structure with BCR more than 1.0 has pay back period less than the life span of the structure. Details of computation of pay back period of farm pond irrigation system are discussed by Mishra et al. (1998).

(iv) Internal Rate of Return

Internal rate of return *(IRR)* is computed as that interest at which the *BCR* value is just 1.0 (von Open and Subba Rao, 1980; Brooks et al., 1990; Islam et al., 1998).

11.7 Economics of Farm Pond

11.7.1 Simulation Study

Economic analysis of farm pond irrigation system based on 30 years of simulated data with farm pond sizes varying from 6 to 20 percent of the farm area (farm area is 800 m^2) for rice-mustard cropping system is shown in Table 11.1 (Panigrahi, 2001). The depth of the ponds for all sizes and side slopes are 2 m and 1:1, respectively. From this table it is observed that at 12 percent size of the farm pond, the net profit, *BCR, IRR* are maximum and *PBP* is minimum and hence this is the optimum size of the pond. The net profits obtained for different sizes of farm pond for 30 years of simulation periods when fitted to normal distribution produced concave shaped curves which are depicted in Fig. 11.1. From this Fig. 11.1, it is found out that as the size of the pond increases, the net profit also increases at any probability of exceedence (PE) levels and at 12 percent size of the pond, net profit is maximum. However, any size more than 12 percent yields proportionately less profit and for the size beyond 20 percent, the profit is negative. Thus, the graphical representation of the economic study of net profit indicates that 12 percent is the optimum pond size for the aforementioned cropping system.

Table 11.1 Average values of economic parameters of farm pond irrigation system for a farm area of 800 m^2 with rice-mustard cropping system.

Pond size (percent farm area)	Total cost, Rs	Present worth of returns, Rs.	Net profit, Rs.	Benefit--cost ratio	Internal rate of return %	Pay back period Years
6	1604.26	1901.26	297.00	1.19	14.4	16
7	1973.35	2351.54	378.19	1.19	14.5	16
8	2335.62	2782.92	447.30	1.19	14.5	16
9	2770.47	3328.38	557.91	1.20	14.6	16
10	3273.17	3919.17	646.00	1.20	14.6	16
11	3671.07	4382.55	711.48	1.19	14.5	16
12	4095.08	5012.43	917.35	1.22	15.0	15
13	4566.10	5119.75	553.65	1.12	13.5	19
14	5024.72	5296.70	271.98	1.05	12.4	23
15	5378.61	5529.17	150.56	1.03	12.1	24
16	5922.58	5721.10	-201. 48	0.96	11.0	30
17	6417.85	5843.00	-574.86	0.91	10.1	30
18	6909.68	5961.00	-948.67	0.86	9.3	65
19	7394.77	6039.02	-1355.76	0.82	8.5	8
20	7876.46	6113.32	-1763.13	0.77	7.7	8

11.7.2 Experimental Study

A field experiment was conducted by Panigrahi (2001) for two years (1999 and 2000) in the experimental farm of the department of Agricultural and Food Engineering at IIT, Kharagpur with 12 percent plastic lined embankment type farm ponds in a farm area of 800 m^2.. Depth of pond being 2 m below ground and pond has a side slope 1:1. The pond has embankment height 30 cm, and the side slope of the pond is 1:1. The berm width of the pond is taken as 30 cm. Pond water serves as supplemental irrigation to rice-mustard cropping system with farm area serving both as catchment and command area. Inflation and the interest rates are assumed as 6 and 12 percent, respectively. The life span of the farm pond is assumed as 25 years. Economic analysis of the experimental data is presented in Table 11.2. Data in Table 11.2 reveals that it is economically feasible to invest in farm pond irrigation system.

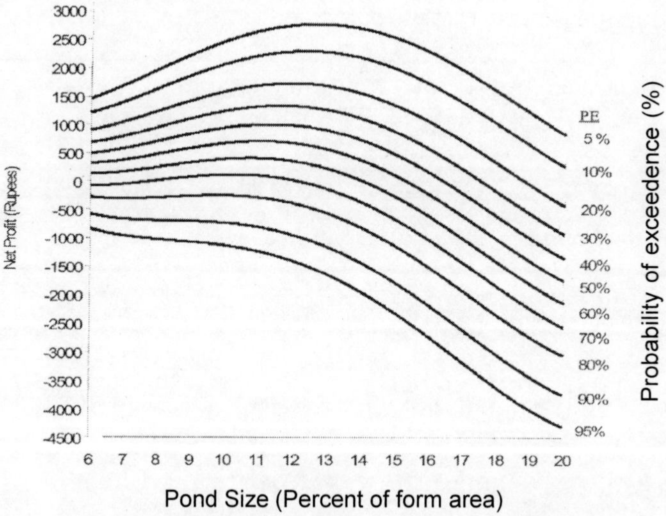

Fig. 11.1 Variation of net profit of farm pond irrigation systems

Table 11.2 Economic feasibility of farm pond irrigation system

Parameters	Year		
	1999	2000	Mean
Initial investment, Rs.	2743.27	2743.27	2743.27
Present worth value of total annual cost, Rs.	1399.39	1092.65	1246.02
Total cost, Rs.	4142.66	3835.92	3989.29
Present worth value of total return, Rs.	5234.02	4143.29	4688.65
Net profit, Rs.	1091.36	307.37	699.40
Benefit-cost ratio	1.26	1.08	1.17
Internal rate of return, %	16.1	13.4	14.8
Pay-back periods, years	13	19	16

In an on-farm water management study by the scientists of Water Management Project of Chiplima, Orissa, three sample unlined farm ponds were taken for study in the Hirakud command for evaluation of economics of the supplemental irrigation to various crops and the results are shown in Table 11.3. The open water surfaces of these three ponds were 728, 1146 and 1106 m^2 and depths were 2.5 m, 3.5 m and

3.3 m, respectively. The results indicate that the ponds are techno-economically viable in rainfed farming systems (Anonymous, 2006).

Table 11.3 Economic analysis of pond irrigation systems in on-farm water management study area in Hirakud command

Economic indicators	Farm ponds			Average
	1	2	3	
Total cost of pond, Rs	2,68,390	4,47,090	3,57,750	3,57,743
Present worth of returns of crops, Rs	3,99,980	6,18,690	4,37,880	4,85,516
Present worth of returns of fish, Rs	33,556	31,850	30,145	31,850
Total present worth of returns, Rs	4,33,536	6,50,540	4,68,025	5,17,367
Net profit, Rs	1,65,146	2,03,450	1,10,275	1,59,623
Benefit -cost ratio	1.62	1.46	1.31	1.46
Pay-back period, years	4	5	6	5
Internal rate of returns, percent	22.0	19.2	17.8	19.7

Question Banks

Q1. What do you mean by economic analysis? Why is it important in determining the optimum size of farm pond?

Q2. What are the main economic indicators to conduct the economic analysis of any project? Define these indicators with relevant formulae.

Q3. Write notes on the following economic indicators:

(i) Net profit (ii) Benefit-cost ratio (iii) Internal rate of return

Q4. What do you mean by present worth analysis? List out its importance in study of the economic analysis.

Q5. What is initial investment and how is it calculated for the farm pond irrigation system?

Q6. How can you compute the total investment for supplying irrigation through a farm pond?

Q7. What is variable cost? How can you estimate the present worth value of the total annual cost or variable cost?

Q8. How is the present worth of annual returns from farm pond irrigation systems calculated? How can it be increased?

Q9. Describe how can the farm pond irrigation system be economically feasible in rainfed faring system

Q10. If the pay back period of a project is less than its life span, is the project economically viable? Justify your answer.

Q11. The initial investment of a farm pond is Rs. 15,600 and the total annual cost is Rs. 540. Total returns obtained from farm pond irrigation system are Rs. 2800. Compute the net profit and benefit-cost ratio of the irrigation sysem and justify whether the system is economically feasible? Assume the interest and inflation rate as 10 and 5 percent, respectively and life span of the structure as 25 yeas.

Q12. With a graph explain how the net profit of farm pond irrigation system varies with the size of the farm pond. How can you find out the optimum size of farm pond from this graph.

Q13. Why do the net profit and benefit-cost ratio decreases for a farm pond irrigation system with larger size beyond the optimum value?

References

Anonymous. 2006. Annual Report of All India Co-Ordinated Research Project on Water Management, Chiplima Centre, Orissa University of Agriculture and Technology, Bhubaneswar.

Brooks, K.N., Gregersen, H.M., Lundgren, A.L and Quinn, R.M 1990. Manual on Watershed Management Project Planning, Monitoring and Evaluation. A Publication of ASEAN-US Watershed Project. College Laguna, Philippines 4031, 274 pp.

Islam, M.T., Saleh, A.F.M. and Bhuiyan, S.I. 1998. Agro-hydrologic and economic analyses of on-farm reservoirs for drought alleviation in rainfed ricelandss of northwest Bangladesh. Rural and Environmental Engineering, Vol. 35: 15-26.

Mishra, P.K., Rama Rao, C.A. and Siva Prasad, S. 1998. Economic evaluation of farm pond in a micro-watershed in semi-arid alfisol deccan plateau. Indian J. Soil Conserv, Vol. 26(1): 59-60.

Palmer, W. L., Barfield, B.J. and Haan, C.T. 1982. Sizing farm reservoirs for supplemental irrigation of corn. Part II. economic analysis. Trans. ASAE, Vol. 15: 377-380, 387.

Panigrahi, B. 2001. Water Balance Simulation for Optimum Design of On-Farm Reservoir in Rainfed Farming System. Ph.D. Thesis, Indian Institute of Technology, Kharagpur.

Panigrahi, B; Panda, S.N; and Mal, B.C. 2007. Rainwater Conservation and Recycling by Optimal Size On-Farm Reservoir. Resources, Conservation and Recycling. (Elsevier Publication), Vol. 50 (4): 459-474.

Taylor, G.A 1975. Managerial and Engineering Economy. D. Van Nostrand Company, second edition, New York.

Thompson, G.T., Spiess, L.B. and Krider, J.N. 1980. Farm resources and system selection. InDesign and Operation of Farm Irrigaion Systems. Ed. M.E. Jensen, ASAE, Monograph 3, St. Joseph, MI: 45-73.

Verma, H.N and Sarma, P.B.S. 1990. Design of storage tanks for water harvesting in rainfed areas. Agric. Water Manage., Vol.18: 195-207.

von Oppen, M and Subba Rao, K.V. 1980. Tank Irrigation in Semi-Arid Tropical India. Progress Report 9, Economics Program, ICRISAT, Patancheru, Andhra Pradesh: 25pp.

Chapter 12

Farm Pond Maintenance

12.1 Introduction

Farm ponds are used for multipurpose including as a viable source for supplemental irrigation to rainfed crops, a good place of entertainment, a vital place for wild life habitats, livestock and fish and aquatic plants. Like any other structure, the farm pond needs regular repair and maintenance for effective operation. The purpose of construction of the pond will be fruitful and the life span of the pond will be enhanced if proper care and maintenance are taken. In this chapter some salient discussions on care and maintenance of the pond are mentioned.

12.2 Maintenance Against Silting

The farm ponds constructed in medium to deep black soils are subjected to getting silted up. Hence, periodic desilting to restore the original storage capacity is required. If the pond is used for supplemental irrigation to crops, then desilting may be done once in every 5 to 10 years depending on the amount of silt getting deposited in the pond.

The breaches and rill formed in the spoil bank and shoulder bunds should be immediately repaired. The inlet and outlet structures should be maintained periodically. The silt deposited in the silt trap should be removed periodically. In areas where silt carried by the runoff is more, a vegetative filter near the inlet structure may be used. The pond may be desilted manually if it is small in size otherwise mechanical measures may be used to desilt. It is easy to desilt the pond when it is dry. A Simple and inexpensive method to desilt the pond is to stir the water of the pond and then allow the desilted water move away from the pond through the outlet (Belgaumi, *et. al.,* 1997). Fig. 12.1 shows a badly maintained pond with excess silting and weed infestation causing severe reduction of its storage capacity.

Fig. 12.1 Schematic view of a silted and weed infested pond

Several substances can be used to hasten the settling of silt particles, including agricultural lime (1000 pounds per surface acre), hydrated lime (1000 pounds per surface acre), aluminum sulfate (250 pounds per surface acre), or agricultural gypsum (300 - 500 resent pounds per surface acre).

Quick lime (calcium oxide) in pond water should not be used. Gypsum and aluminum sulfate may increase the acidity of the pond water, rendering the pond less suitable for fish. To reduce the risk of harming aquatic organisms, one may apply only one-fourth of the recommended amounts of chemicals /aquatic herbicides at a time and wait for about two weeks to assess whether the desired effect can be achieved at a lower rate. One can make the most of an application by mixing the substance with clear water to form slurry, which can then be sprayed and mixed evenly across the surface of the pond. This will ensure a uniform mixing of the herbicides in pond water and the concentration of the substance in water of the pond will be evenly distributed.

12.3 Pond Maintenance for Wildlife Habitats and Livestock

The maintenance of farm pond depends on the purpose for which it is potentially used. Most of the ponds are vitally important for wildlife but because we see them every day in our life, their importance is overlooked and undervalued. Many rare and familiar plants and animals depend on ponds for at least part of their life cycle. They provide breeding areas for frogs and toads, shelter for water voles and otters and feeding areas for bats and birds. Recent research has shown that ponds in agricultural landscapes often have more freshwater wildlife than rivers, streams or ditches. Moreover many farmers have farm ponds that serve as a watering source for livestock and as a fishery for themselves. Farm pond management, if not taken seriously, can have severe health concerns for the livestock.

There is no standard type of pond that is best for wildlife and livestock but there are a few general things that can help make any pond a rich wildlife habitat. The following tips will help maintain the pond to enrich wildlife habitats and livestock (http://pond.dnrcornell.edu)

- Don't clear all the trees around ponds. Partial shade is important for some rare plants and animals and may help to control some problem plants.
- Always plan any work to take account of what is already there -clearing and dredging isn't always the best thing to do.
- Clearance or dredging (if it is needed) at any time of the year will affect some species. So only do it if really necessary and at a time when it will do least damage.
- Do not introduce any alien species to the pond.
- Avoid polluting the water in the pond. The essential root of the health problems that arise in farm ponds is animal waste. Farm ponds that experience high levels of animal waste will also have low levels of oxygen, due to the amount of dissolved oxygen required to degrade the waste. This imbalance can lead to the production of noxious and toxic gases. The gases that are normally produced in these circumstances are ammonia and hydrogen sulfide, both of which can cause pneumonia and other illnesses in livestock. In addition to the production of harmful gases, the lack of dissolved oxygen can also result in a high mortality rate

for the fish that reside in the pond as well as leading to an increase in algae and other unwanted aquatic plant growth.

- Avoid fertilizer and pesticide contamination in pond water. When using any fertiliser or pesticides near ponds follow the appropriate codes of good practice and avoid pesticide drift over ponds.
- Make sure the runoff areas feeding the pond are free from chemicals. If runoff is inevitable, build a smaller settling pond upstream from the main pond to catch and retain nutrients
- Don't apply manure or slurry within 10 metres of a pond.
- Leave a buffer zone around a pond whenever possible to provide habitat and a physical barrier against disturbance and spray drift.
- Falling leaves from nearby trees and bushes, as well as leaves from pond plants, can quickly build up on the bottom of the pond. Settling to the bottom, the decaying vegetation can increase levels of dissolved organic compounds (DOCs) and rob water of oxygen, stressing fish in the pond that are already struggling to deal with fall temperature swings. During late summer and early fall it is important to remove as much material from the bottom of the pond as possible. Drape a pond net over the pond to prevent the majority of leaves and twigs from getting into the water in the first place.

12.4 Aquatic Weed Control in Farm Ponds

Aquatic plants in the farm ponds provide many benefits, including prime fish habitat, shading, erosion control, oxygenation, and scenic beauty. The owners of the farm ponds should become familiar with the plants in their ponds and determine if or when a plant is overabundant. If plants are interfering with common uses of a pond (fishing, swimming, irrigation), then they should be identified and treated accordingly.

Many methods of aquatic weed control are now available. Before applying to any particular treatment, one has to locate the particular contaminated source and access the magnitude of contamination. One has to determine if excess nutrients are entering the pond from a septic area, agricultural operation, or treated lawn. Substances like nitrogen and phosphorus create algae blooms and excessive plant growth. Nutrient sources should be controlled before implementing aquatic weed treatment.

Floating weeds like duckweed and algae masses can be skimmed off the surface with a skimmer or rake. Hand-pulling of root systems, and grass carp can be used to reduce submerged plant growth. If the weed growth in the pond is maximum, then the ponds can be drained partially in the winter to kill aquatic weeds near the shoreline without disrupting fish and wildlife. If required, the water of the weed infested pond may be discharge and harvested in an auxiliary adjacent pond and then the weed infested pond after being dried may be dug out to remove all the silts and weeds. Then the water harvested in the auxiliary pond may be rediverted to it. Some pond owners seek quick remedies for pond plant problems. Aquatic herbicides are available for this purpose; however, pond owners should consider herbicides a last resort. The disadvantages of chemical treatments include high cost, risk of deoxygenation, the need to obtain permits, impacts to fish and swimmers, and limitations based on pond characteristics.

12.5 Farm Pond Maintenance Routines

A farm pond, no matter how well built, requires sufficient maintenance in order to meet the farm operator's objectives. Ponds constructed for fishing, swimming, irrigation, wildlife, or fire suppression all require regular inspection for emerging trouble. Like other farm structures, farm ponds serve an important function for some farm operations and require a higher level of maintenance. Pond maintenance means inspecting different areas of a pond on a regular basis for signs of erosion, water quality decline, wildlife damage, and aquatic weed control. The pond needs regular maintenance in the structural parts like inlet and outlet structures, maintaining stability of pond sides etc.

12.5.1 Maintenance of Spillways

The most significant feature of a well-constructed pond is a proper spillway - the point at which excess water exits a pond. All ponds require some type of spillway. But there are instances that many ponds fail each year because of insufficient spillways. Newer ponds use pipes to collect water and discharge it at the base of the dike. Spillways and emergency spillways should always be kept clean and free of woody vegetation. If the spillway is clogged, heavy rains in the monsoon season can cause big problems (Shah, 1986).

12.5.2 Pond Bank Erosion

Erosion of soil in or near a pond can create a very serious hazard for pond owners and neighbors downhill. An unexpectedly heavy rain or spring thaw can intensify a small erosion problem without warning. Steep shores without vegetation can erode to cause cloudy water conditions and pond silting. Pond owners should check the dike and sides of ponds carefully several times a year. One has to check especially for unusually soggy ground near the base of the dike, muddy water seeping through the dike, rills (long channels in the soil) and soil slumping down the dike.

Grass helps prevent erosion on the exposed areas near a pond. Fill eroded areas and re-seed or place sod over bare soil. The roots will retain and stabilize the soil. The interior of dikes can erode if loose material or wood has been used as fill to create the dike. Also, water can seep along the outside of a spillway pipe, carrying away soil and creating a hazard.

The farm pond should be protected against soil erosion from the banks and sides occurring from rainfall, wave action, livestock etc. which damage to the embankment. Most of the exposed surfaces should be protected by grass sodding or with grass cover. The upstream side of the embankment should made stone ripraping for protection against the action of wave where the pond is expected for a wide fluctuation and should be provided from the top to the bottom of the embankment. The stones should be durable and large in size which can not be displaced by the wave action. Berms with sliping inward are provided at normal pond level with grass turfing. The berm should be of 2.5 to 3.0 m wide (Murty, 1998).

12.5.3 Protection of Dikes/Embankments of Pond

Newly built dikes of the pond should be protected against erosion by planting or seeding a grass cover on the crest of the dikes and also on both sides i.e. on their dry side and wet side down to the normal water level of the pond. Sometimes the seeds may be sown on the loosen soil of the dike and if required light irrigation may be provided for germination. The following tips will help to have a quick spread of the grass cover in the dike (Coche et al., 1995; Shah, 1986).

(i) Spread a 10-15 cm layer of topsoil over the area to be planted. The top soil is either brought back to the pond area from which it was earlier removed or is obtained from a nearby source.

(ii) If the soil moisture is not optimum for seed germination or stress condition occurs for the growth of the seedlings, then provision of light watering may be done.

(iii) Sprinkler irrigation may be used to irrigate the soil sown with grass seeds or planted with grass seedlings. Sprinkler irrigation does not create soil erosion.

(iv) For quick growth of grass, fertilizers may be applied. For example 13-13-13 mixture (NPK) at the rate of 50-100 gm/m^2 surface area.

(v) When it rains heavily, temporary protection measures such as hay or other suitable materials may be adopted to avoid rill erosion of the pond sides and dikes.

(vi) Mowing of the grass should be done at regular intervals. Mowing will help a quick and uniform establishment of grass cover.

(vii) Big plants should not be allowed to grow on or near the sides of the ponds. The roots of the plants will weaken the dikes. If required, vegetable crops and forage bushes can be grown but care should be taken to select plants with a good ground cover with shallow roots so that the roots do not disturb the soil of the dikes.

(viii) Grazing on the dikes should not be allowed.

(ix) The pond should be protected through barbed wire or live fencing in order to prevent the human beings and other animals from slipping and falling in to the pond and destroying the lining materials.

12.5.4 Cares Taken While Lining Pond

Some ponds especially ponds in coarse textured soil, lining is required. A cost effective lining material is to be selected. Further, the longevity of the lining materials needs to be taken into consideration before selecting a particular lining material. Sometimes, locally available materials are given preference for lining. If pond is lined by plastic films, the following cares must be taken for increasing the life span and effective storage of the pond (Panigrahi, 2001).

(i) After the pond is constructed, steps of at least 30 cm width and 30 cm height must be made forming a series of bench steps.

(ii) The gravels/pebbles, roots of the plants etc. must be taken out from the soil on the steps as well as from the bottoms of the pond.

(iii) Plastic films are then laid over the steps. There should not have any cut or holes in the plastic films and the gaps in the joining pieces of the plastic films must be perfectly sealed. It is preferable to use a single plastic sheet for lining the farm pond so that the no joining of the cut pieces are required.

(iv) Gently put soil (without any gravels/pebbles etc.) over the steps and gently compact it. Fill the soil and compact in gradual steps till a gentle slope of the pond side is maintained.

(v) Do not allow the cattle or any other animals to enter the pond or graze on the pond sides to avoid tearing of the plastic films by the roofs of the animals.

(vi) Cares must be taken while fishing in the pond so that the plastic films are not destroyed.

(vii) Change the plastic films and relined the pond as discussed above every 5-10 years interval.

(viii) If a pump is used to irrigate the crops by using the pond water, then cares must be taken so that the plastic films are not destroyed by the operation of the pump.

(ix) While desilting the pond, cares must be taken so that the films are not cut to pieces.

12.6 Farm Pond Maintenance Checklist

Although the different maintenance measures of the farm pond have been discussed in details in the foregoing sections, the following checklist will help in easy and systematic maintenance of the ponds (http://pond.dnrcornell.edu).

After Storms or Heavy Rain

- ❒ Inspect the dikes of the pond for erosion.
- ❒ Inspect quality of runoff entering the farm pond and reduce nutrient input if present.

Monthly

- ❒ Assess presence of aquatic weeds in the ponds; excessive growth typically occurs in mid to late summer.

- ❒ Mow the grasses regularly to stabilize the dikes; remove woody vegetation from dike.
- ❒ Check water temperature for fish habitat.

Seasonally

- ❒ Inspect for wildlife damage.
- ❒ Repair or replace safety signs.
- ❒ Catch and release fish to determine population health.
- ❒ Inspect spillway for stability.
- ❒ Inspect fencing to restrict livestock from entering water.

Annually

- ❒ Assess silting or changes in pond depth.
- ❒ Record maximum / minimum depths.
- ❒ Check water pH and adjust for fish habitat.

Question Banks

Q1. How does the farm pond get silted up?

Q2. What is desilting? Why is it important to desilt the farm pond periodically?

Q3. How can you desilt small as well as large farm ponds?

Q4. Describe the various methods used to desilt a farm pond.

Q5. Name some materials used to hasten the settling of silt particles in a farm pond.

Q6. Why gypsum and aluminum sulfate are not recommended for use in fish farm pond?

Q7. What is the procedure to use the chemical / aquatic herbicides in farm pond without harming the aquatic organisms?

Q8. How are the ponds useful for wildlife habitats and livestock?

Q9. Mention the different steps required to be undertaken to maintain the pond to enrich wildlife habitats and livestock.

Q10. What are the harmful affects of contaminated water in the farm pond?

Q11. What are the sources of pollution of water in the farm pond?

Q12. Mention the various methods to be employed to control the growth of aquatic plants in farm pond.

Q13. What are the advantages and disadvantages of using chemicals or aquatic herbicides to kill the weeds in farm pond?

Q14. How can the spillways of farm ponds be maintained well?

Q15. What do you mean by pond bank erosion? How can it be controlled?

Q16. How do the grasses help in checking the soil erosion of the bank of the farm pond?

Q17. What do you mean by stone ripraping? How can it help in controlling the soil erosion of the sides of the farm pond?

Q18. Briefly describe the various ways by which the dikes of the farm ponds can be protected.

Q19. What steps are to be followed to have quick and uniform grass coverage on the dikes of the farm ponds?

Q20. Write the different cares to be taken for lining the farm ponds with plastic sheets.

Q21. Why should a farm pond be protected by barbed wire or live fencing?

Q22. What is a farm pond maintenance check list? List out the various measures to be taken up on monthly, seasonal as well as annual basis.

References

Belgaumi, M.I., Itnal, C.J. and Radder, C.D. 1997. Farm Ponds. Technical Series 4, Publication Centre, University of Agricultural Sciences, Dharwad, pp. 47.

Coche, A.G., Muir, J.F. and Laughlin, T. 1995. Simple Methods for Aquaculture Pond Construction for Freshwater Fish Culture Building Ponds, FAO of the United Nations, Rome, pp. 355.

http://pond.dnrcornell.edu. Farm pond safety and responsibility.

Murty, V.V.N. 1998. Land and Water Management Engineering. 2nd Edition, Kalyani Publishers, New Delhi, pp. 586.

Panigrahi, B. 2001. Water Balance Simulation for Optimum Design of On-Farm Reservoir in Rainfed Farming System. Ph.D. Thesis, Indian Institute of Technology, Kharagpur.

Shah, S.L. 1986. Methodology, programme policies and economics of watershed management in the Himalayan region of India. Proceedings of seminar on "Environmental Security and Watershed Management-Relevance of People's Participation" held at Pantnagar.

Chapter 13

Other Sources of Irrigation

In addition to farm ponds, irrigation to crops can be met from other sources. The main sources can be broadly classified as: (i) Canal irrigation, (ii) Well Irrigation and (iii) Lift irrigation. In arid and semi arid areas, where availability of water is scarce, pitcher irrigation is practiced. These sources of irrigation are briefly mentioned as follows.

13.1 Canal Irrigation

To feed the world population of 8 billion in 2025 AD, it is necessary to bring more area under irrigation because irrigated farms typically get higher yield under provision of irrigation. The food production will have to increase by about 50 percent to feed some two billion more people by 2025 AD, and a large part of that increase needs to come from irrigated agriculture. It is reported that the yield of different crops are generally 2-3 times higher than the crops grown under rainfed conditions. Furthermore, it is possible to raise 2-3 crops round the year under irrigated agriculture whereas hardly a single crop is successfully grown in a year under rainfed condition. It is reported that about 40% of the world's food production comes from only 17% of the irrigated land of the world (Sivanappan, 2000). Everybody is aware of the fact how the irrigation played a crucial role to make the Green Revolution in India successful. A major part of irrigation is achieved by canal irrigation through different irrigation schemes.

A project that makes use of weir or a barrage for providing irrigation to fields is called as direct irrigation scheme. On the other hand, a storage irrigation scheme makes use of storage reservoir by construction of dams. In order to provide irrigation by both the schemes, it is necessary to have a network of canal systems including main canals, branch canals, distributaries, minors and watercourses. It is imperative to design the canal systems properly for a certain realistic value of peak discharge that must pass through them so as to provide sufficient irrigation to the command areas. These canals have to be aligned and excavated either in alluvial soils or non-alluvial soils; depending upon which they are called alluvial canal or non-alluvial canal.

13.1.1 Alluvial and Non-alluvial Canals

(i) Alluvial Canals

The soil which is formed by continuous silt deposition is called alluvial soil. The alluvial soil is even and is having a flat surface slope. Hard foundations are generally not available in this kind of soil. The rivers flowing through such alluvial areas, have a tendency to shift their courses. The river bed consists of sand of considerable thickness, and is, therefore, permeable. When ever, an irrigation structure is to be constructed on such a river, special precautions and design method is to be adopted. The canals when excavated through such soils are called alluvial canals. Canal Irrigation is generally preferred in such areas as compared to storage irrigation.

(ii) Non-Alluvial Canals

Mountainous and plateau regions may suffer from weathering. Weathering over a period of time result in the formation of a rocky plain area, called non-alluvial area. The rivers passing through such areas have no tendency to shift their courses and they do not pose many problems for designing irrigation structures on them. Canals, passing through such areas are called non-alluvial canals. Storage irrigation is preferred to canal irrigation in this type of areas.

13.1.2 Alignment of Canals

Irrigation canals can be aligned in any of the three ways as:

(i) Watershed canal

(ii) Contour canal

(iii) Side-slope canal

(i) Watershed Canal

The dividing line between the catchment area of two drains is called the watershed (Fig. 13.1).

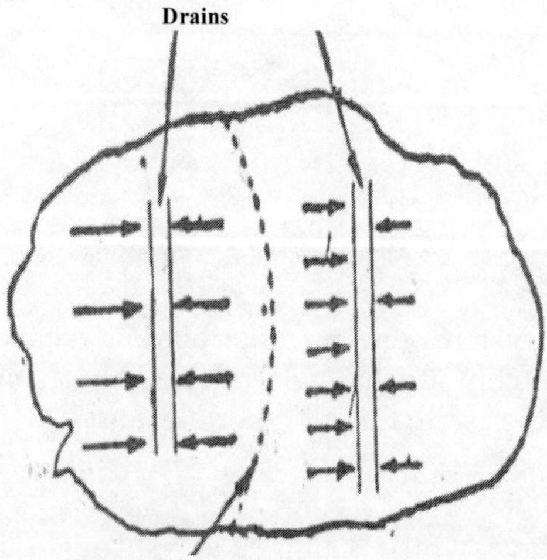

Fig.13.1 : Alignment of a watershed canal along the watershed

For canal system in plain areas, it is advantageous to align all channels on the watersheds of the areas. The canal which is aligned along any natural watershed is called a watershed canal. In such a canal, water flows by gravity, on either side of the canal, directly or through small irrigation channels. More, cross-drainage works are avoided as the natural drainage will never cross a watershed because all the drainage flows away from the watershed.

(ii) Contour Canal

In hilly areas, the river flows in the valley, while the watershed or the ridge line may be hundred of metres above it. It becomes uneconomical to take the canal on top of such a ridge. The channel, in such cases, is generally aligned parallel to the area, except that the longitudinal slopes required to generate sufficient flow velocities, are given to it (Fig. 13.2).

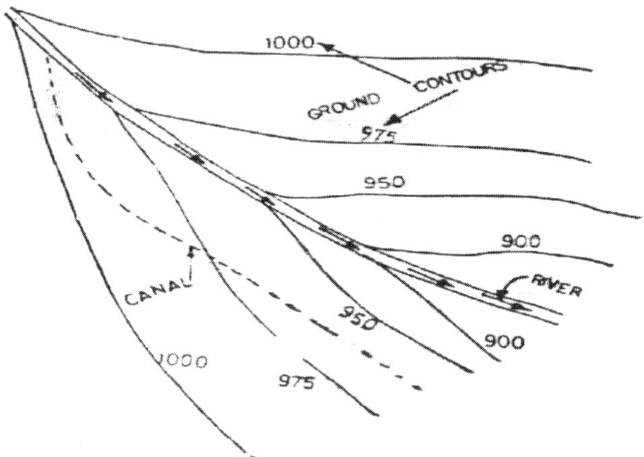

Fig. 13.2 : Alignment of a contour canal

Contour canal system requires canal falls and cross drainage works at suitable places. A contour channel irrigates only on one side, because the areas on the other side are at higher elevation.

(iii) Side Slope Canal

A side slope canal, which is aligned at right angles to the contours, i.e along the side slopes. Such a channel is parallel to the natural drainage flow. It doesn't intercept cross drainage, and hence no cross drainage works are required.

13.1.3 Distribution System for Canal Irrigation

Irrigation water has to be supplied to the crop fields from the head regulator constructed over a storage reservoir or weir/barrage. The entire network of irrigation channels is called Canal System. The canal system as explained consists of:

(i) Main canal
(ii) Branch canal
(iii) Distributaries
(iv) Minors and
(v) Watercourses

In case of direct irrigation scheme, a weir or a barrage is constructed across the river, and water is stored on up-stream side. The arrangement is known as head works or diversion head works. Water is diverted into the main canal by means of a diversion weir. A head regulator is provided at the head of the main canal, so as to regulate the flow of water into the main canal. In storage irrigation scheme, a dam is constructed across the river, thus forming a reservoir on the upstream side of the river. The water from this reservoir is taken into the main canal through the outlet sluices. There are generally two main canals, which off-take from the reservoir, called Left Bank Canal and Right Bank Canal, as shown in Fig. 13.3.

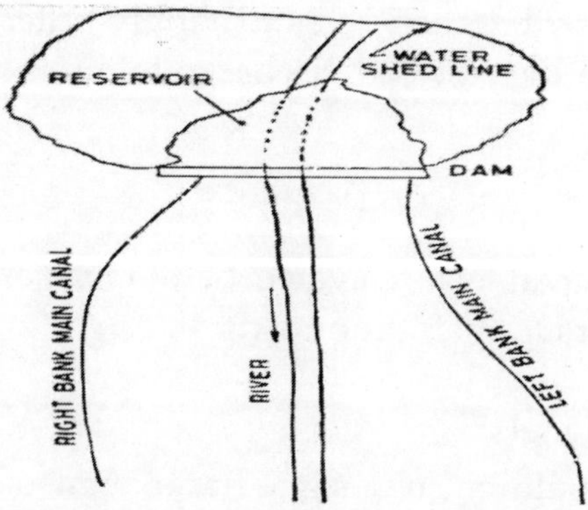

Fig. 13.3 : Storage irrigation scheme

(i) Main Canal (Head Reach)

The canal head works are generally situated in a valley, and the canal should mount the watershed in the shortest possible distance. The canal, in this reach, must be aligned very carefully and has to be generally excavated in deep cutting below natural ground surface level. Sometimes, it has to cross various drainage lines. In the portion below the head reach of the main canal, attempts are made to align the canal along the watershed and somewhat central to the commanded area. Sometimes, watershed has to be sacrificed to bypass town, villages etc. that come on the way.

(ii) Branch Canal

The canals that originate from the main canal and take out water from it and supply to the smaller channels called as distributaries are called as branch canals.

(iii) Distributaries

Smaller channels which take off from the branch canals and distribute their supply through outlets into minors or watercourses, are called distributaries. They are aligned either as watershed channels or as side slope channels. Discharge in a distributary, is generally, less than 30 cumecs.

(iv) Minors

Sometimes, the fields in a command area are situated in far away distances such that the distance between the distributary outlet and field is very large. In such a case, small channels called minors are taken off from the distributaries, so as to supply water to the farmers at the point nearer to their fields. Discharge in a minor, is generally, less than 2.5 cumecs.

(v) Watercourses

These are small channels excavated by the farmers, so as to take water from the minors to their crop fields.

13.1.4 Curves in Channels

The channels should be so designed and aligned that is has less curves. But many a times the curves become inescapable. However, if curves in a channel are unavoidable, every care should be taken so that the curves have gentle slope. A curve cause disturbance of flow and results in silting on the inside (i.e convex side) and scouring on the outside (i.e concave side) (Fig. 13.4). Pitching is, therefore, sometimes proposed on the concave side so as to avoid scouring. If the discharge is more, the curve should be more gentle and should, therefore, have more radius. The generally adopted minimum values of radii for different discharges are tabulated in Table 13.1 (Garg, 1987).

13.1.5 Certain Important Definitions

Gross Commanded Area (GCA)

It is the total area, bounded within the irrigation boundary of a project. The boundary of GCA is usually defined by the drainage in either side across which irrigation cannot be expanded economically.

Table 13.1 : Minimum recommended values of curve radii of channel for different discharges

Discharging capacity of channels, cumecs	Minimum recommended curve radius, m
< 0.5	100
0.5-3	150
3-15	300
15-30	600
30-100	900
> 100	1500

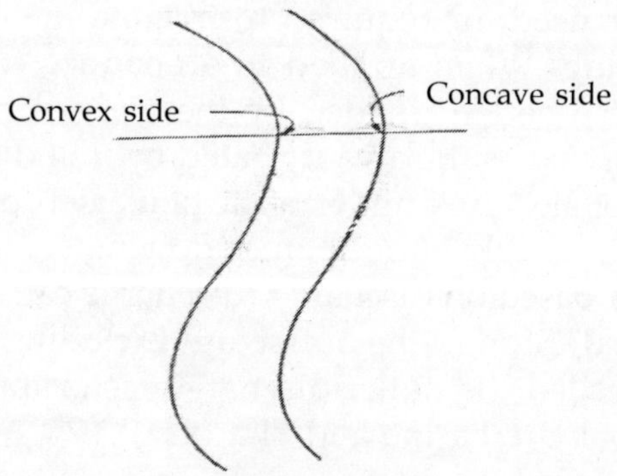

Fig. 13.4 : Slope alignment in channel

Cultivable Command Area (CCA)

It is that part of the gross command area in which cultivation is possible. It is less than the gross command area. The area under small drainage, ponds, abadies, reserve forest roads, railways, hills etc. is excluded from the gross command area to get the cultivable/culturable command area.

Intensity of Irrigation

It is the percentage of CCA which is irrigated annually or seasonally. It varies with the type of irrigation, climate of the area, water management practice of the crops, rainfall and its distributions etc.

Area to be Irrigated

It is the area to which irrigation is provided. It is to be remembered that not all the CCA are the irrigated areas.

Crop Period

It is the time taken by a crop from sowing/planting till harvesting. The crop period varies with the types of crops and also for the same crop, it varies with the type of varieties.

Base Period

It is the time period between the first watering of a crop at the time of sowing to its last watering before harvest.

Duty

It is the relation between the area irrigated, or to be irrigated, and the amount of water used, or required to irrigate the crop for the purpose of its growing. When applied to a channel, it is the area irrigated during a base period divided by the mean supply used in cumecs. It may be defined as the area irrigated by unit discharge that means the number of hectares under a particular crop brought to maturity by a constant supply of one cumec flowing continuously for the base period. Duty based on discharge (discharge calculated at the head regulator) passed through the outlet and excluding all losses in the canal system is called the outlet discharge factor and is usually represented by symbol *D (*Sharme and Sharma, 2002*)*.

Delta

It is an expression used in the irrigation practice to mean the depth of water that would result over a given area for a given discharge for a certain length of time. It is a term equivalent to duty of irrigated water when the latter is expressed in water depth units. Delta is the total depth of water in cm needed by a crop for its growth. The average delta for rice (short duration) is 80-100 cm, for medium and long duration rice it is 100-120 cm and 120-150 cm, respectively. The delta of non-rice crops are generally less: pulses have values about 30 cm, oil seed crops, 40 cm and for most of the vegetables it is 25-35 cm. The delta of crops depends on the types of crops, crop varieties, duration of the crops, and climate of the areas where the crops are grown.

Relationship between Duty and Delta

Let D be the duty of water in hectares/cumecs; B be the base period in days and Δ be the delta of water in metre.

Volume of one cubic metre of water flowing in one day

$= 1 \times 24 \times 60 \times 60 = 86,400 \text{ m}^2 = 8.64$ ha m.

Volume of one cubic metre flowing for B days = 8.64 B ha m.

By definition of duty D, one cubic metre = 10,000 D m^2 of water supplied for B days matures D hectares of land.

Total depth of water applied in the land = volume/area = 86400 B/10000 D. = 8.64 B/D m.

Now delta is the total depth of water, Δ = 8.64 B/D m which gives

$$D = 8.64\,B/\Delta \text{ m and } \Delta = 8.64\,B/D \text{ m} = 864\,B/D \text{ cm.}$$

13.1.6 Discharge Computation at the Head Regulator

Discharge can be computed by average outlet factor as D = 864 B? where D is duty in hectares/cumec, B is in days and ? is delta of water in cm. However, a channel designed on this average outlet factor will prove to very in adequate, as it will fail to supply the required water to the crop at its peak demand. It is recommended that a channel be designed to carry maximum discharge which is estimated by taking into account the peak water demand in the peak period which will be clear in some of the worked out examples as mentioned below. Provision of channel losses should be there to decide the final channel capacity.

To obtain precise channel design capacity, one should find out the monthly or 15 days water requirement of various crops. The water depth required in this interval is multiplied by the crop area so as to give us the volume of water required in this interval. Dividing the volume by interval, we can find out the discharge required in each interval, by various crops; the summation of which for all the crops will give us the discharge required by all the crops in each interval. The channel may then be designed, for maximum of these values. It is evident that in order to be more precise, the interval should be as small as possible. Generally, monthly water requirement studies are conducted, and the channel capacity is increased by 20 to 25% to cater for peak demand in the month (Garg, 1987).

Example 1. The gross commanded area for a distributary is 5000 hectares, 60% of which is cultivable. The intensity of irrigation for *rabi* season is 50% and that for *kharif* season is 30%. If the average duty at the head of the distributary is 1800 hectares/cumec for *rabi* season and 1000 hectares/cumec for *kharif* season, find out the discharge required at the head of the distributary from average demand considerations.

Solution

G.C.A = 5000 hectares

C.C.A = 5000 x 60/100 = 3000 hectares.

Area to be irrigated in *rabi* season

= C.C.A x intensity of Irrigation

= 3000 x 50/100 = 1500 hectares

Area to be irrigated in *kharif* season

= 3000 x 30/100 =900 hectares.

Water required at the head of the distributary to irrigate *rabi* area

= 1500/1800 cumecs = 0.833 cumecs

Water required at the head of the distributary to irrigate *kharif* area,

= 900/1000 = 0.900 cumecs

Thus, the requirement in *kharif* season is 0.900 cumecs and that in *rabi* season is 0.833 cumecs. The required discharge is maximum of the two. i.e. 0.900 cumecs.

Hence, the distributary should be designed for 0.900 cumecs discharge at its head from average demand considerations.

Example 2. The gross command area for a distributary is 5000 hectares, 60% of which is cultivable. The intensity of irrigation for *rabi* season is 50% and that for *kharif* season is 30%. Wheat is grown in the entire command in *rabi* and paddy in *kharif*. Find out the discharge required at the head of the distributary considering the peak water requirement of *rabi* crop (wheat) 15 cm and the period of peak water requirement 4 weeks and that for *kharif* cop (paddy) 30 cm with peak period of water requirement 4 weeks.

Solution

Outlet factor for *rabi* crop (wheat) = 864 x B/D= 864 x (4x7)/15.0

= 1613 hectares/cumecs and that for *kharif* crop (paddy)

= 864 x (4 x 7)/30

= 806 hectares/cumecs

Area to be irrigated in *rabi* season

= C.C.A x intensity of Irrigation

= 3000 x 50/100 = 1500 hectares

Area to be irrigated in *kharif* season

= 3000 x 30/100 =900 hectares.

Water required at the head of the distributary to irrigate *rabi* area

= 1500/1613 = 0.93 cumecs

Water required at the head of the distributary to irrigate *kharif* area

= 900/806 = 1.12 cumecs

The required discharge is maximum of the two i.e.1.12 cumecs.

Note: The required discharge from peak or maximum water requirement consideration have gone up to 1.12 cumecs from 0.90 cumecs.

Example 3. The cultivable commanded area of watercourse is 1000 hectares. Intensities of irrigation of sugarcane and groundnut crops (*rabi* season crop) are 30% and 40% respectively. The duties for the crops at the head of the watercourse are 750 hectares/cumec and 1400 hectares/cumec, respectively. Compute the discharge required at the head of the watercourse.

Solution

Cultivable command area. = 1000 hectares

Intensity of irrigation for sugarcane = 30%

Area to be irrigated under sugarcane

= 1000 x 30/100 = 300 hectares.

Intensity of irrigation for groundnut = 40 %

Area to be irrigated under groundnut

= 1000 x 40/100 = 400 hectares.

Given duty for sugarcane = 750 hectares/cumec

Duty for groundnut = 1400 hectares/ cumec

Hence discharge required for sugarcane = 300/750 cumec

= 0.40 cumec and discharge required for groundnut

= 400/1400cumec = 0.285 cumec

Now, sugarcane requires water throughout the year and groundnut for only rabi season. Hence the water requirement at the head of the watercourse at any time of the year will be the summation of the two, i.e. equal to 0.40 + 0.285 = 0.685 cumecs.

It is to be noted that the discharge during rabi season will be 0.685 cumec and for the rest of the year, it will be 0.40 cumec.

13.1.7 Design of Irrigation Channels

Water carried by the irrigation canals contains a lot of derbies, silts and many other foreign materials. Often, the alluvial canals carry a lot of alluviums which get deposited in course of time in the channel beds and side. Water moving with a given velocity and a certain depth can carry in suspension, only a certain amount of silt of a certain nature. If water of a given velocity and depth is not fully charged with silt (that it can carry in suspension) it will scour the bed and sides of the channel, till it is fully charged with silt. Hence, if the velocity of flow in the channel is more, the bed and the banks are likely to the eroded, and similarly, if the velocity is less, the silt which was formerly carried in suspension is likely to be dropped. Silting and scouring in channels is very common and must be avoided by proper designs. Scouring lowers the full supply level and causes loss of command area. It may also cause breaching of canal banks and failure of foundations of irrigation structures. Silting interferes with the proper working of a channel, as the channel section gets reduced by silting, thereby reducing its discharging capacity. Therefore, while thinking to design a proper functioning channel, one must think to design such a channel in which neither silting nor scouring takes place. Such channels are known as stable channels or regime channels.

Regime Channels

A channel is said to be in a state of 'regime', if the flow is such that 'sitting and scouring' need no special attention. Such a state is not easily possible in rivers, but in artificial channels, such a state can be obtained by properly designing the channel. Design of regime channels is based on mainly two approaches: (i) Kennedy's theory and (ii) Lace's theory.

Kennedy's Theory

R.G. Kennedy, an Executive Engineer of Punjab P.W.D. carried out extensive investigations on some of the canal reaches in the upper

Bari Doab Canal System. He selected some straight reaches of the canal section, which had not posed any silting and scouring problems during the previous 30 years or so. From the observations, he concluded that the silt supporting power in a channel cross-section was mainly dependent upon the generation of the eddies, rising to the surface. These eddies are generated due to the friction of the flowing water with the channel surface. The vertical component of these eddies try to move the sediment up, while the weight of the sediment tries to bring it down, thus keeping the sediment in suspension. So if the velocity is sufficient to generate these eddies, so as to keep the sediment just in suspension, silting will be avoided. Based upon this concept, he defined the critical velocity (V_0) in a channel as the mean velocity which will just keep the channel free from silting or scouring and related it to the depth of flow by the equation.

$$V_0 = c\, D^{n} \quad (13.1)$$

where c and n are constants depending upon silt charge.

Values of c and n were found to be 0.55 and 0.64, respectively.

Therefore, Eq. (13.1) can be written as

$$V_0 = 0.55\, D^{0.64} \quad (13.2)$$

But the formula (Eq.13.2) is true only for the Upper Bari Doab Canal System and not for any other reaches. Later realizing this lacuna in the above equation, he introduced a factor depending upon the type of soil. This factor is named as critical velocity ratio (C.V.R.) (denoted by symbol m) which depends upon the silt grade. The modified Kenned's formula is given as (Sharma, 1987).

$$V_0 = 0.55 \text{ m } D^{0\text{-}64} \quad (13.3)$$

where, V_0 = Critical velocity in the channel in m/sec and D = Depth of flow in the channel, m.

For sands coarser than the standard, the values of m are taken from 1.0 to 1.2, and for sands finer than the standard, m is taken as a value between 0.9 to 0.8 as shown in Table 13.2 (Sharma and Sharma, 2002).

Table 13.2 : Recommended values of critical velocity ratio, m for different silt carried channel.

S. No.	Type of silt	Value of m
1.	Silt of river Indus (Pakistan)	0.70
2.	Light sandy silt in North Indian River	1.0
3.	Light sandy silt, a little coarser	1.1
4.	Sandy, loamy silt	1.2
5.	Debris of hard soil	1.3

Design Procedure: The design procedure is based on the following steps.

i. Assume a trial depth, D and then compute the critical velocity V_0 by the above Eq (13.3).

ii. Determine area by dividing discharge by velocity.

iii. Then determine channel dimensions.

iv. Finally, compute the actual mean velocity (V) that will prevail in the channel of this cross-section, by using Kutter's formula, Manning's formula, etc.

v. If the two velocities V_0 and V works out to be the same, then the assumed depth is all right, otherwise change it and repeat the procedure. Thus several iterations are required to get the dimensions of the channel.

There are a number of formulae used to compute the flow velocity. The two main formulae are:

Kutter's Formula :

$$V = \left[\frac{1/n + \left(23 + 0.00155/S\right)}{1 + \left(23 + 0.00155/S\right) \dfrac{N}{\sqrt{R}}} \right] \sqrt{R\,S} \tag{13.4}$$

Manning's Formula :

$$V = \frac{1}{n} R^{2/3} S^{1/2} \tag{13.5}$$

where V = Velocity of flow, m/s;

R = Mean hydraulic radius which is area/wetted perimeter, m;

S = Bed slope of the channel, m/m and

n = Rugosity coefficient.

The values of "n" in both these equations depend upon channel condition, and also upon discharge. The values of "n" may be taken as given in Table 13.3 (Sharma and Sharma, 2002).

Table 13.3 : Recommended values of Rugosity coefficient for unlined channels

Condition of channel	Value of n
Very good	0.0225
Good	0.025
Indifferent	0.0275
Poor	0.030

Example 4. An irrigation channel has side slope 1:1 and base width 3 m. It has flow depth of 1 m. The bed slope of the channel is 0.04 percent. Assume Manning's rugosity coefficient n = 0.028. Determine if the channel will be regime channel. Assume critical velocity ratio as 1.0.

Solution

Given n = 0.028 and S = 0.04 percent = 1/2500

Base width = 3 m and side slope is 1:. The depth of flow = 1m.

Hence top width of the channel = 5 m.

Area of the trapezoidal section = (5 +3)/2 = 4 sq. m.

Wetted perimeter = 3 + 2 $\sqrt{2}$ = 5.83 m.

Mean hydraulic radius = A/P = 0.69 m.

Using the above values of n, R and S, velocity by Manning's formula, V = 0.56 m/s.

Using Kennedy's formula, velocity V_0 = 0.55 m D $^{0\text{-}64}$ = 0.55 x 1.0 x. $(1.0)^{0.64}$ = 0.55 m/s.

Since the two velocities are nearly same, the channel is stable i.e. regime.

Lacey's Theory

According to Lacey's theory, there is a relationship between velocity, V and mean hydraulic radius, R in the form V = $R^{1/2}$, As a measure of silt grade, Lacey introduced a silt factor, f and he proposed an equation (Sharma and Sharma, 2002) as:

$$V = K\,(f\,.R)^{1/2} \tag{13.6}$$

where K is a constant. The value of K proposed is $K = (2/5)^{1/2}$

Hence Eq. (13.6) can be written as:

$$V = (2/5)^{1/2}\,.\,(f.\,R)^{1/2} \tag{13.7}$$

Hence, $f = 5/2.\ V^2/R$ (13.8)

The value of f depends on average size of boundary materials in the channel and is given by the equation:

$$f = 1.76\ m_r^{1/2} \tag{13.9}$$

where m_r is average particle size, mm.

13.1.8 Lining of Irrigation Canals

India stands second next to China in having maximum irrigated area. Huge amount of money is spent in all most all the Plan periods for irrigation infrastructure development. Irrigation water is a costly commodity, and as such, there should be no wastage during its carriage from the sources or reservoirs to the fields. Every care should be taken so that the losses in the channels including application losses are kept minimum so that maximum area is irrigated. Most of the canals, constructed in India to carry irrigation water, are unlined, and hence, a large part of the costly irrigation water is lost in seepage and percolation. In some regions the seepage losses are very small, whereas in some regions, upto 50% of the water is lost in seepage, thus, those areas are required to be lined. Seepage loss is a very serious problem which reduces the irrigation potential of the supplied water, which necessitates lining of irrigation channels. Fig. 13.5 shows the photographic view of an unlined channel in Hirakud canal command which shows that there is heavy seepage loss consequently resulting in low discharge in the canal. On the other hand Fig. 13.6 shows the photographic view of a concrete lined channel which represents that the seepage losses in the channel is very less which consequently increases the channel carrying capacity.

Fig. 13.5 : Photographic view of an unlined channel in Hirakud canal command

Fig. 13.6 : Photographic view of a concrete channel of Hirakud canal system

13.1.9 Advantages of Lining

The advantages of lining of irrigation canals are described below in short (Michael, 1978):

(i) Seepage Control

The seepage losses are considerably reduced if the channels are lined. A lined channel costs about 2-2.5 times as much as an unlined channel, but where seepage is very large, the saving of costly irrigation water may itself be sufficient to fully justify the capital expenditure on lining. The heavy seepage losses in channels would necessitate the construction of larger reservoirs and bigger dams. Prevention of seepage by lining would, therefore, reduce their impounding capacity, and hence, lower the construction costs of these works. The data taken for seepage study in a part of Hirakud canal system both before and after lining are presented in Table 13.4 (Anonymous, 1994).

Table 13.4 : Rate of seepage in cumecs per sq. m (x 10^{-6}) of wetted surface area in unlined and lined channels.

S.No.	Types of lining	Rate of seepage in cumecs per sq. m. of wetted area before lining (x 10^{-6})
1.	Unlined channels	4.1
2.	30 cm ×15 cm × 5 cm tiles using cement mortar 1:2 (1 cement: 2 sand)	0.15
3.	7.5 cm thick cement concrete 1:3:6 (1 cement: 3 sand: 6 brick ballast)	0.14
4.	7.5 cm thick cement lime concrete 1:5:10:20 (1 cement: 5 lime: 10 surkhi: 20 brick ballast)	0.43

(ii) Prevention of Waterlogging and Salinity

Heavy seepage losses through unlined channels, often causes the water-table to rise in the surrounding fields upto or near to the ground level. The rise in water table often bring the crop /plant roots within the capillary fringe. This, in turn, brings up the alkali salts near to the ground surface, which soon renders the land to be salined making it unfit for cultivation. Such a land is usually called *thur,* and the phenomenon of rise of water table is known as water-logging. It is well-known that large tracts of lands in Punjab (India and Pakistan) have been rendered uncultivable in this manner. Thus, lining of canals prevents seepage and thus protects cultivable land. Combined with land drainage schemes, lining helps to reclaim water-logged areas.

(iii) Increase in Channel Capacity

The capacity of a channel can be considerably increased by lining it, due to the fact that, the lining presents a smooth surface and, therefore, causes less resistance to the flow of water. The water, therefore, flows faster, and hence, more of it is carried per second than that in an unlined channel. Flow of water in unlined channels, is often, further retarded by the vegetation on the sides and bottom and since capacity is a function of velocity, the higher the velocity, the greater is the capacity of the channel.

(iv) Increase in Command Area

A lined channel can be designed not only smaller in cross-section but also shorter in length. The steeper gradients can be provided because higher velocities are permissible (as the material is less erodible) and a shorter alignment can, therefore, be selected. On the other hand, flatter slopes can be provided without silting on a lined channel compared to these on an unlined channel. It can, therefore, help to bring high areas under command.

(v) Reduction in Maintenance Costs

The keep-up of unlined channel involves huge sum of recurring expenditure, generally, charged under the head of account: annual repair and maintenance of the canal system. This expenditure may be required on:

i. Periodical removal of silt;

ii. Minor repairs;

iii. Removal of weeds and water plants.

(vi) Elimination of Flood Dangers

An unlined channel constructed on weaker foundations is always a danger, and a breach may occur at any time. Instances have occurred where small breaches in unlined channels resulted in washing away of considerable length of embankment- leading to flooding of certain areas and causing scarcity of irrigation waters in other areas, as the canal remain out of service at the critical time of crops.

13.1.10 Construction of Various Types of Linings

The construction practices for the various types of lining are discussed as follows. In general, rough excavations can be economically carried out with some excavating machines, and fine grading may be done by manual labour. Various types of channel linings, which are commonly used are (Michael et al., 1970; Sharma and Sharma, 2002):

(a) Hard Surface or Rigid Linings

Various types of rigid linings are described below.

(i) Cement concrete lining;

(ii) Plaster lining;

(iii) Brick tile or concrete tile lining;

(iv) Asphaltic concrete lining.

(b) Earth Type Linings

There are two types of earth type linings. They are:

(i) Compacted earth;

(ii) Soil cement.

Hard Surface or Rigid Linings

(i) Cement Concrete Lining

This type of lining made from selected aggregates is considered as a high-quality type of lining. They give very satisfactory service. Despite the fact that they are frequently higher in initial cost, their long life and minimum maintenance usually make them most economical. Concrete linings have been extensively used in most of the developed countries. India being a developing nation, finds it difficult, to afford their high initial costs and as such, their use is restricted only to important projects. It is also not feasible from financial point of view to have cement concrete lining in the entire canal system. Only the main canal and branch canals at the head regulator and the bends in canal systems are lined with cement concrete.

(ii) Plaster Lining

Plaster is a paste of cement mortar applied under pressure through a nozzle on the surface of the channel. It consists of a mixture of cement and sand (generally in the ratio of 1:4). Larger proportions of cement are required in plaster as compared to what is required in cement concrete. Wire mesh reinforcement is generally, although not necessarily, used in plaster linings. The plaster lining is of immense use for smaller jobs, because of the lighter equipment and small crew required to carry out the work. It is also useful for repair works and in rehabilitation of old canals. Plaster linings can also be placed in an irregular canal section, thus eliminating the necessity of trimming the section, as is required for concrete canal linings when placed with a slip-from. This type of lining is generally laid in a thickness of about 3.5 cm.

(iii) Brick Tile or Concrete Tile Lining

This type of lining is very popular in India because of certain inherent advantages in their use. These advantages are:

- Brick or concrete tiles can be laid by ordinary masons and no skilled labour is required.
- Rigid quality control is not required.
- No expansion joints are required.
- Rounded sections can be easily laid without using any formwork.
- Larger number of skilled and unskilled labour is required, thus, providing greater employment potentials.

The specifications may provide either a single or a double layer of brick masonry or tiles laid in mortar. Sometimes a layer of tiles is laid over a layer of brick masonry. The top layer of tiles is generally laid in 1:3 cement mortar over 15 mm thick layer of plaster in 1:3 cement plaster.

(iv) Asphaltic Concrete Lining

This material is used as lining material at very few places. It is still under the stage of evolution. Asphaltic concrete is a carefully controlled mixture of asphalt and graded stone aggregate mixed and placed under elevated temperature. It provides a fairly cheap lining, especially where the asphalt is available within the country. It is flexible and readily confirms to the sub-grade. The disadvantages or the limitations of this type are as follows:

- It does not decrease the rugosity coefficient of the channel.
- It permits certain type of weed growth.

Earth Type Linings

(i) Compacted Earth Lining

Soil graded to obtain the required characteristics and containing enough fines, so as to make it impervious, is thoroughly compacted at optimum moisture content, and is used to provide a lining of 30 to 90 cm thickness. The use of this type of lining is restricted to the availability of suitable soils in the area, through which the canal is being constructed. It would be uneconomical to transport the selected materials from outside.

(ii) Soil Cement Lining

Portland cement up to the extent of 2 to 8 % is added to the soil having a high percentage of fineness. The mixture is as first mixed dry. Water is then added so as to bring the soil to its optimum moisture content, and again mixed thoroughly. Material is then placed at the site and is compacted. Curing is then required for at least 7 days by covering with wet sand.

13.1.11 Maintenance of Irrigation Channels

Irrigation channels are nothing but earthwork constructions, and as such, very much susceptible to damage. They therefore require a lot of maintenance; upkeep watches and word, etc. as to ensure their continuous efficient functioning. Various problems which are posed by the irrigation channels during their use and need regular repair and maintenance are (Michael and Ojha, 1966):

1. Silting of channels,
2. Weed and plant growth,
3. Failure of weaker banks,
4. Hollows created by burrowing animals, crabs, etc. in the field up sections causing seepage and piping,
5. Channel breaches due to piping, overflow of channel.

Some of these factors, such as the first and second reduce the efficiency of the canals, whereas the others may result in complete seepage of canal supplies to the irrigated fields. All these causes of trouble and their remedial measures are discussed below.

13.1.11.1. Silting of Channel

Irrigation channels normally get silted during their course of flow. Whenever the flow velocity in the channel reduces, the silt carried by the water in suspension gets deposited on the bed and sides of the channels. The silt so deposited reduces the effective channel cross-section which consequently reduces the carrying capacity of the channel. In order to prevent too much of silt deposition, irrigation channels must be properly designed, so as to ensure a velocity which neither causes any silting nor scouring in the channels. This necessitates the construction of regime channels. However even in such canals the discharge may sometimes be less than the designed and in that eventually silting may take place in the channels.

Attempts should also be made to remove the silt from the water entering the channels itself. From the headwork, for this purpose, almost silt free water must be admitted in the channel. However in spite of using all these measures, certain amount of silt will definitely enter the channel and certain amount may be added to it, by some scouring from within the channel. This silt often gets deposited on the bed and sides of the channel.

The deposited silt normally deforms the shape of the channel, and reduces its carrying capacity. It must, therefore be removed practically by desilting operations. The desilting can be carried out by actually excavating the silt either manually or by dredgers. However, if canal functions properly and is in regime and taking its full supply, it is not necessary to clear the silt to the theoretical cross-section. But if the canal is not functioning properly, desilting is required, but only to extent of of clearing a portion of it as to get the canal back to efficient working order.

13.1.11.2. Weed and Plant Growth

Irrigation canals normally get infested with aquatic weed growth. The silted canals provide excellent base for weed growth. The sun rays also help in promoting weed growth. The weed growth hinders the flow of water, reduces the carrying capacities of the channels, and thereby impairing their efficiencies. The weed growth may also sometime impart harmful qualities to irrigation water.

Weed growths may be controlled by maintaining higher velocities in the canals. Moreover, many a times, the weed growths will have to be physically removed from the canals in order to improve their efficiencies. In case of newly constructed canals, regular inspection should be undertaken to locate any spots where the weed growth has set in. weed from such spots should be removed complete, so that the infestation does not spread. In the case of old canals, where aquatic weed growth is profuse, suitable mechanical or chemical methods may be employed, at as early stage as possible.

13.1.11.3. Failure of Weaker Banks

The banks of the canal should be made strong, and should be maintained regularly. Due to constant use by men and animals, and movement of vehicles, the canal banks get eroded and breaches occur at various places. They must be repaired and all cuts and breaches filled up with suitable soil and proper tamping. Moreover, grass or turfing should not be scrapped while smoothening the bank surfaces, as it helps in stabilizing the soil and thus preventing its erosion. However, long grasses will have to be cut as far as necessary to smoothen the surface of the bank and to avoid holes being hidden under high grasses. Sometimes when a canal reach runs in filling, the banks are subjected to water pressure. In such a case there exists every possibility of damage occurring to the banks. In such cases, the banks must be given extra strength by increasing their sections. In order to keep their cost of construction low, additional soil for increasing the bank section may be obtained by the natural silting process.

13.1.11.4. Canal Breaches

Canal breach is, in fact, an opening or a gap developed in the canal bank due to erosion of some portion of the bank. When once water starts flowing out of such a breach, it start becoming deeper and wider, unless remedial measures are taken immediately to plug the breach. The breaches in canal banks may be caused due to various reasons, such as:

a. Breach due to faulty design or construction of the canal banks,

b. Breach due to overflow of the canal,

c. Breach due to seepage or piping, and

d. Breach due to intentional cuts made by cultivators.

13.2. Well Irrigation

In India, ground water withdrawal from wells for irrigation has been practiced since long and it is still now an important source of irrigation. Most of the wells draws supplies from the first water bearing structure and discharge about 5 litres/sec (lps) depending on the static water table position and geological formations.

13.2.1. Advantage of Well Irrigation

1. It is owned by the farmers and as such has full control on sources of irrigation. It is also a handy source of irrigation.
2. It is located nearer to the place where the irrigation water has to be served to crops and so conveyance loss is minimum.
3. Particularly suitable for intensive irrigation to raise cash crops and is suitable to grow 2 to 3 crops from the same piece of land in a year.
4. Low cost of construction.
5. Reliable source of irrigation.
6. It can be used as a conjunctive source with canal irrigation
7. Supply is more uniform in quality and chemical composition and as such more helpful to crop growth and yield.
8. Supply is free from weeds, turbidity and bacterial pollution which is associated with canal water
9. Economical water use having higher delta and low duty.
10. Benefit of rainfall can be availed. If there is chance of rainfall then well irrigation can be withheld.

13.2.2 Disadvantages of Well Irrigation

1. Associated higher repair with maintenance and hence higher operating cost.
2. It has low discharge and consequently small area is irrigated by it.
3. Water is debris/silt free. So more fertilizer is required to be added to crop field to raise crop yields.
4. If the velocity of water from well exceeds critical velocity of water in the soil that convey water to well, then there is chance of well subsidence.

13.2.3 Location of Well

1. Site of open well should be in the central position of the farm so that water can be served to all parts of farms easily without more conveyance loss.

2. It should be located at a higher elevation so that water can flow by gravity to all parts of farms.
3. Site should have good water bearing aquifer to supply sufficient yield to well.
4. If there exists any well, then the near well should be at least 75 - 175 m away from it to avoid any interference of wells and avoid low yield
5. It should be very much away from any source of pollution
6. There should not have any adjoining boundary lines within the circle of reference of the well.

13.2.4 Open Wells

It is essentially a vertical hole of diameter 2 to 3 inch i.e bigger than that of tubewell having less depth compared to tubewell and hence is associated with less yield. The yield of open well is limited to 5 litres/sec. It draws water from any one pervious stratum. However, it is recommended that the well be constructed to a depth below ground water table such that the well does not dry during summer. From financial point of view, the depth is limited to 30 m. Contribution of water to well is from both bottom and sides of wells in unlined case. If well is lined, contribution is only from bottom.

13.2.5. Types of Open Wells

Open wells may be dug, bored, driven or drilled type. Dug wells are constructed manually or by semi mechanical means. Bored wells are constructed by hand or by power augers. Drilled wells are made constructed either by percussion or rotary drills and the excavated materials are brought to surface by bailer, sand pump, suction bucket or by hydrostatic pressure. Driven wells are constructed by driving a casing at the lower end. Open well can be further classified as unlined well or lined well. It may also be shallow or deep well.

13.2.5.1 Unlined Wells

It is also called as kacha well having no lining either at bottom or at sides.. These wells have depth 3 - 3.5 m. They are constructed in stable pervious soil so that side subsidence will not occur and groundwater can be easily tapped. These wells are cheap and easy to construct. But their longevity is less since there is chance of side collapse. The flow enters the well both from sides and bottom. Fig. 13.7 represents the view of an unlined well.

13.2.5.2 Lined Wells

Depending on the types of lining materials used, lined wells may be pervious lined type or impervious lined type.

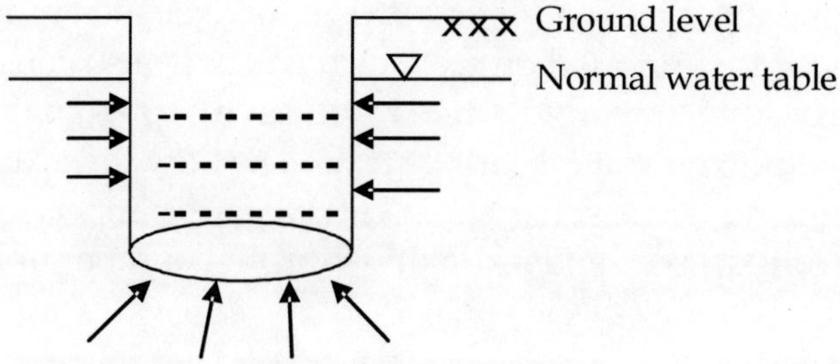

Fig. 13.7 : Unlined well

13.2.5.3. Pervious Lined Wells

Such wells have dry bricks or stone lining with open joints through which water enters the well. In sandy formation in order to avoid flow of sand surrounding the lining into the well, the space behind the lining is packed with brick ballast of size 20-25 mm. These wells are comparatively small depth and are suitable for shallow water table and for gravel and sand formation. The flow of water into the well is both from bottom and sides. Flow is of both radial and spherical pattern. Wherever pervious wells are plugged at bottom, flow pattern is radial type only. Fig, 13.8 gives the view of pervious lined wells.

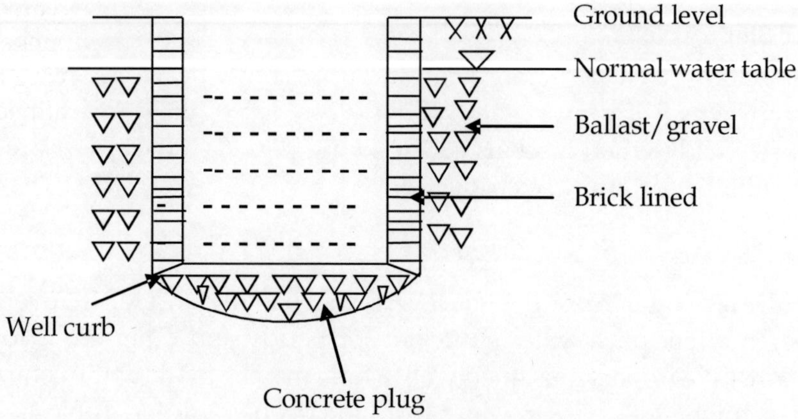

Fig. 13.8 : Pervious lined well

13.2.5.4. Impervious Lined Wells

In non alluvial soil, if the soil is sandy or soft and pervious, then there is chance of side collapse. So lining of sides of wells is necessary. Such wells need an impervious lining with brick masonry or stone masonry in cement or lime mortar or with reinforced concrete. Steining of 30-60 cm thick is supported on well curb which is also constructed of reinforced concrete. Impervious lined well have depth ranging from 5 to 30 m depending on water bearing strata. The flow towards well is through bottom and hence spherical. A better type of impervious well draws water from deeper sandy formation through a pipe sunk down from the centre of the well (Fig. 13.9).

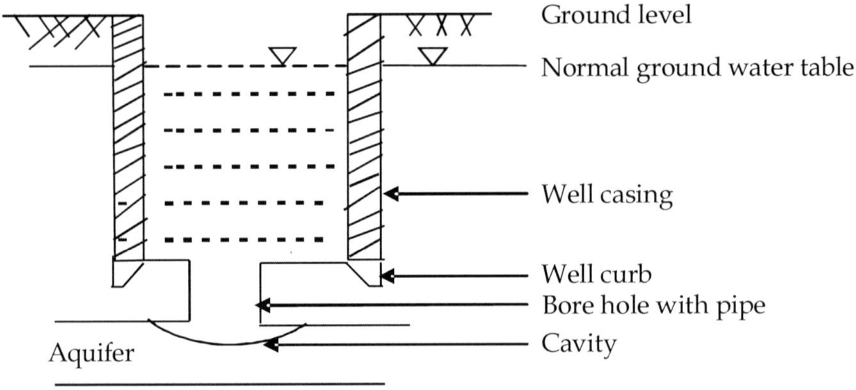

Fig.13.9 : Impervious lined well

13.2.5.5 Shallow Well

A shallow well penetrates into the uppermost pervious stratum only from which it draws water. It is of less depth; yield is less and dries up quickly when discharged for a long period. It cover less area and is suitable for non rice crops.

13.2.5.6 Deep Well

It draws water from deeper layers and has high yield. It covers more area. It supplies water sufficient for paddy and other crops for irrigation. Like impervious lined well, a hole may be made at the centre of the well which draws up water from the deeper aquifer. Table 13.5 gives difference between the shallow and deep wells.

Table 13.5 : Difference between shallow wells and deep well.

S. No.	Shallow well	Deep well
(i)	Initial cost low	Initial cost high
(ii)	Cross sectional area at bottom of well less which limits high	Cross sectional area high due to cavity formation supply to well and so supply more water to well
(iii)	Suitable where ground water table is very near to the ground	Suitable even if ground water table is deep below surface ground surface.
(iv)	Yield is less	Yield is high
(v)	Quality of water is not good	Better quality water

13.2.6 Yield of an Open Well

Safe yield of a well is the yield taken out as discharge from a well without causing any mining. It is expressed as m^3/ hr or lit / min. Under stable condition, water level in the well coincides with that in the ground water table in the adjoining aquifer. But as the pumping in the well continues, water level in the pumping well falls immediately where- as the levels in adjoining aquifer remains at higher elevation. This forms a cone of depression around the well. Difference between the water level in the well and that in the aquifer is known as draw down or depression head (Slichter, 1989). The more the rate of withdrawal, the more is the depression head. The depression head is maximum at the centre of the well and gradually decreases as the radial distance from well increases. At radius of influence, it is zero. There are two tests for determining the yield of an open well. They are pumping test and recuperation test.

13.2.6.1 Pumping Test

It is an accurate method to determine the safe yield of an open well. In this method, water level in the well is decreased through heavy withdrawal till maximum safe depression head is reached. At this stage, the pump speed is regulated so that the rates of withdrawal equals the recharge. Thus constant discharge water level is maintained in the well which creates equilibrium condition. The ratio of pump discharge per hour under this equilibrium condition gives the yield of the wells at a particular draw-down

Applying Darcy's law

$$V = K\,i = K.\ h/r \tag{13.10}$$

where V= velocity of flow of water into the well at equilibrium condition, i = Hydraulic gradient equal to ratio of drawdown (h) to radial distance (r) from wells and K= constant called as hydraulic conductivity of the aquifer.

Discharge (Q) is given as:

$$Q = A.V = A.\ K.\ h/r \tag{13.11}$$

Hence, considering unit width aquifer we get $A = r$

$$Q = r.\ K.\ h / r$$

$$Q = K.h \tag{13.12}$$

The yield of a well can also be expressed in another way.

Equation (13.12) can be written as:

$$Q = (K/A)\ A.\ h \tag{13.13}$$

where K / A is expressed as specific yield of the well (m^3 /hr) per m^2 of the aquifer through which water enters into the well from aquifer through unit depression head. The values of K/A for different soils are given in Table 13.7 (Sharma, 1987).

Table 13.7 : Values of K/A for different types of soils

Soil type	Value of K/A, (m^3/ hr per m^2/ unit depression)
Clay	0.25
Fine sand	0.5
Coarse sand	1.0

Example 5. Design an open well in fine sandy soil with safe yield 0.5 m^3/hr /m^2/ unit depression head so that an yield of 0.004 m^3/sec is discharged from well with depression head h = 4 m

Solution

Using Eq (13.13) we have $Q = (K/A).\ A.\ h$

Q = 0.5 (m^3 /hr /m^2 / m depression) X A X 4 m

= 2 A m^3 /hr/m^2

Given Q= 0.004 m^3 /sec = 14.4 m^3 /hr

So 14.4 m^3 /hr = 2A m^3 /hr /m^2

A = 7.2 m^2

Area A = (π/4) . d^2 where d = diameter of well

(π/4) d^2 = 7.2 m^2

$$d = \frac{\sqrt{4 \times 7.4}}{\sqrt{3.142}}$$

d = 3 m.

13.2.6.2 Recuperation Test

It is used where it is difficult to maintain constant water level in the well. In this method the water level is decreased to any level below the normal level by pumping and then pumping is stopped. When pumping is stopped then water level in the well starts rising (Fig. 13.10). Time taken by water in well to attain any draw down below ground water table normal position is recorded. Then the well discharge formula (Eq 13.14) as given below is used.

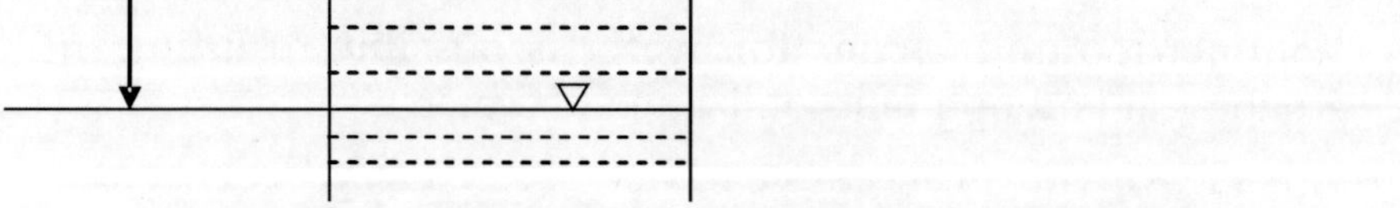

Fig. 13.10 : Yield of an open well by recuperation test

$$Q = \frac{2.3}{t} \log_{10} \frac{h_1}{h_2} A.h \qquad (13.14)$$

where t= time taken by water to attain drawdown below normal ground water table position, h_1 and h_2 are initial and final ground water table position, h = average depression head and A= area of cavity of well bottom through which water enters the well given as:

$$A = \frac{4}{3}\left(\frac{\pi}{4}\right) d^2 \qquad (13.15)$$

where d = diameter of well

Example 6. Calculate the average yield of a well of 3 m diameter by recuperation test. The water level in well is decreased to an extent of 2.5 m. The rate of rise of water is 1.0 m/hr. Assume average depression head, h =3 m

Solution

Given h_1 = 2.5 m

After one hour t= 1 hr, h_2 = 1.5 m

A= (4/3).(π/4). 3^2 and h =3 m

$$Q = \frac{2.3}{t}\log_{10}\left(\frac{2.5}{1.5}\right)\times\frac{4}{3}\times\frac{\pi}{4}\times 9\times 3$$

$= 9 \times 3.142 \times 2.3 \log_{10} (1.67)$

$= 26 \text{ m}^3 / \text{hr} \approx 7 \text{ lit} / \text{sec}$

13.2.7 Tubewell Irrigation

13.2.7.1 Advantage of Tubewell Irrigation

Advantages of tubewell irrigation are :

(i) It is an independent source of irrigation which a farmer can at his convenience dig and maintain the well.

(ii) Some patches of lands which cannot be easily irrigated by other sources can be irrigated by it.

(iii) Volumetric rate of irrigation water is possible resulting in economical and optimum utilization of water.

(iv) Water loss in conveyance system is less since conveyance channels are generally lined.

(v) Since seepage and conveyance losses are minimum there is less water logging problems.

(vi) Irrigation capital investment is less compared to canal irrigation.

(vii) There is quick return from well irrigation. Canal irrigation takes 5-6 years to supply irrigation water where as tube well irrigation is possible in the same year where tube well is setup.

(viii) Optimum utilization of rain water is possible.

(ix) Water is suitable for domestic use as quality of water is good.

13.2.7.2 Disadvantages of Tubewell Irrigation

Following are the disadvantages of tubewell irrigation.

(i) Life of tubewell is limited to 10-15 years

(ii) Operating cost is more and so costlier than canal water

(iii) Cover less area compared to canal irrigation

(iv) Frequent repair and maintenance is required.

(v) Not well suited in areas where elevation is not available since water moves by gravity.

(vi) Tube well water carries less silts and hence crop yield is less. In order to increase crop yields, crops needs more fertilizers

(vii) Chances of theft of well components are more.

(viii) Not suitable for high water requiring crops.

13.2.7.3 Suitability of an Area for Tubewell Irrigation

Tube well irrigation is suitable where the sub soil formation is good for storing water and the aquifer is suitable to supply ample ground water to the well. Water bearing stratum may be confined or unconfined. If aquifer has impervious layer at the bottom and top is open to atmosphere, then it is unconfined. A confined aquifer is confined between two impervious layers. The best place for a tube well digging is sandy or sandy aquifer. If the subsoil strata has a 30-50 m thick coarse sand stratum, then it is ideal to dig well. The tube well has a low yield capacity generally 0.042 m^3 / sec., i.e 42 lps which can generally command irrigation to 160 ha of non paddy crop at 100% irrigation intensity after taking care of rainfall. A deep tube well can supply irrigation water 300 ha of non paddy crop. While selecting a site for tubewell, care should be taken so that it is dug in a high elevation to facilitate in gravity supply of water. It should also be located in the central position of the farm so that water can easily be conveyed to all parts of the farm with minimum conveyance loss. In addition to above mentioned points, liological test, aquifer test and location of aquifer boundary should also be taken into account before selection of site of tube well construction (Ahrens, 1958; Benison, 1947; Dylla and Muckel, 1967).

13.2.7.4 Types of Tubewells

For irrigation point of view tubewells are of two types (i) Direct irrigation tube well and (ii) Augmentation tube well.

(i) Direct Irrigation Tubewell

This is the tubewell which provides irrigation directly to crops in the farm. It may be classified as shallow or deep tube well.

(ii) Augmentation Tubewell

In rainy season, canals runs almost full and there is no shortage of supply to crops. But in winter / summer the canal supply is reduced. At this time the ground water can be pumped in conjuction with canal water. Tube wells are used to feed to canal supply which can be used for irrigation. The augmented tube well can serve for augmentation of canal water supply or as a direct source for irrigation. It can also be used as a viable source for ground water recharge. Tube wells can also be classified as shallow in deep tubewells.

(iii) Shallow Tubewells

As the name suggests, these are shallow in depth, usually they tap water from a few layered aquifers. They have diameter usually same from top to bottom. They cover less irrigation areas as their discharge capacity is less.

(iv) Deep Tubewells

These well are deeper in depth penetrating several multilayered aquifers. They draw water from deep confined aquifers. As augmentation tube well, they can be installed in adjacent to rivers. They are spaced at about 500 m from one another. For a given drawdown the deep tubewell yields more discharge than a shallow tubewell (Fig. 13.11).

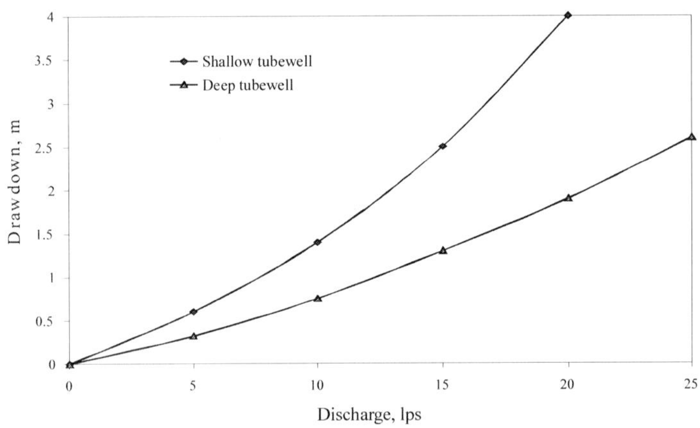

Fig. 13.11 : Discharge drawdown curves of shallow and deep tubewells

13.2.7.5 Hydraulics of Tubewells

A knowledge of well hydraulics essential for the proper design and construction of a tubewell. Well hydraulics is governed by Darcy's law. When water is pumped by a well, water level in the pumping well is lowered down. Thus lowering of water level at any point in the aquifer is termed as drawdown or depression head. The drawdown varies with the distance from the well. It is minimum at centre of tube well and decreases as the distance from tube well decreases forming a cone of depression. Draw down is zero at certain distance at which draw down of tube well is zero. This distance at which draw down of tube well is zero is called as radius of influence, "R". Values of R depends on the well discharge, aquifer materials, time of pumping etc. In general value of R is maximum for a confined aquifer and minimum for an unconfined aquifer (Campbell and Lehr, 1973).

The lowering of water level in tube well causes head difference towards the well which induces water flow into the well from the surrounding water bearing formations of aquifers. The velocity of flow gradually increases towards the well. With increasing velocity, the hydraulic gradient increases as the flow converges towards the well, as a result the lowered water surface has a steeper slope towards the wells. The form of the surface resembles cone shaped depression. The main equation that governs the flow path in a tubewell is generally given by Darcy's law which states that velocity of flow through a porous media (V) is directly proportional to hydraulic gradient (i) i.e $V \alpha i$, or $V=K.\, i$, where K is hydraulic conductivity of aquifer. Since $Q = AV$, we can write $Q = A.\, K\, i$

Velocity of flow through a soil depends on soil type, arrangement of grains and viscosity of water. Darcy's law is valid for laminar flow for which Reynolds number is = 1. It should be remembers that at the vicinity of well, the flow has high velocity and high hydraulic gradient and Reynolds number is > 1, i.e flow is not laminar but turbulent. Under such case Darcy's law cannot be used.

13.2.7.6 Design of Tubewells

Good design of a tubewell aims to ensure an optimum consideration of performances, long service life and reasonable installation and operation cost. The tubewells should give highest yield from the aquifer and should give optimum efficiency in terms of specific capacity. Various design parameters of a tubewell/water well are

(i) Well structure (ii) Well diameter (iii) Well depth (iv) Length of well screen (v) Well screen slot opening (vi) Well screen diameter (vii) Gravel pack design. Gravel pack design is the most important part of design of tube well. Gravel pack design ensures stability of well so that side collapse as tubewell boring is avoided. At the same time, it filters water from aquifer so that good quality water is obtained from the well.

(i) Well Structure

It consists of two main elements (i) the upper portion that houses the pumping equipments and the vertical section through which water flows upward from the aquifer to the level where it enters the pump and (ii) lower intake portion of the well. In lower intake portion, a well screen is used which allows water to enter into the well at low velocity, prevents sand and other fine materials and prevents caving-in of the loose formation materials. The intake portion consists of a bore hole drilled into the aquifer to an adequate depth. Fig. 13.12 gives view of well structure.

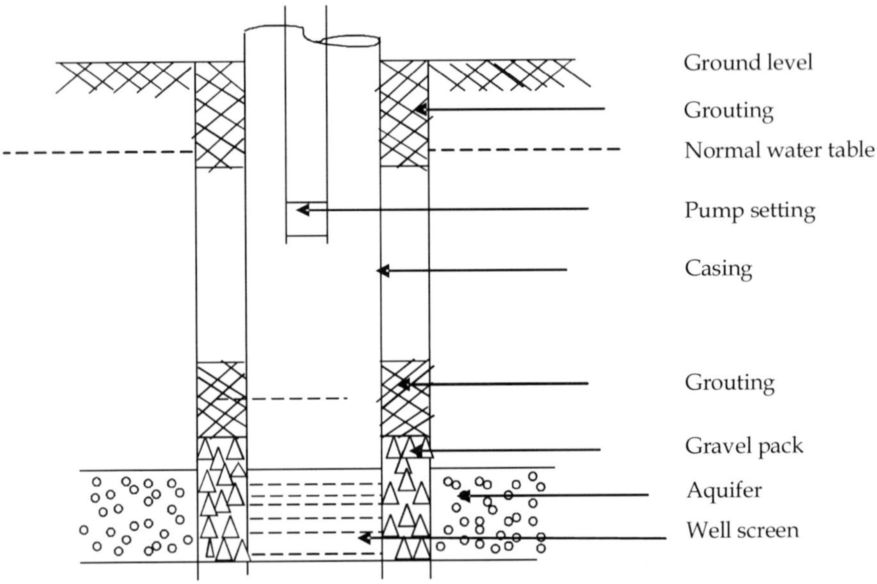

Fig. 13.12 : View of well structure

(ii) Well Diameter

Well diameter is generally governed by the size of the pump. Diameter of the well should satisfy the following condition.

i. Opening should be large enough to accommodate pump and pipes so that it will be easy for installation and efficient operation. The opening of the well should be such that after inserting the casing and housing pipe, 2-3 cm annular space is left between the well and bore hole for gravel packing.

ii. Diameter of the intake section should be large enough as to ensure good hydraulic efficiency of the well. It is not necessary to have a uniform size of well diameter. In deep wells, well diameter below lowest anticipated pump setting, may be reduced to effect some economy. The lowest anticipated pump setting should be carefully estimated keeping in view future decision of ground water table. Table 13.8 (Das Gupta, 1987) gives the recommended size of well casing for different well yields.

The diameter should be so chosen that it gives desired percentage of open area in the screen (15-18%) so that entrance velocity near the screen is within 3 to 6 cm / sec. This will reduce the well losses and hence draw down. Table 13.9 gives representative open area of well screen for various screen sizes (Das Gupta, 1987).

Table 13.8 : Recommending well casing for different wells yields

Anticipated well yield, lpm	Nominal size of pump bowl, cm	Size of well casing Minimum (cm)	Optimum (cm)
Up to 400	10	12.5	15
400-600	12.5	15	20
600-1400	15	20	25
1400-2200	20	25	30
2200-3000	25	30	35
3000-4500	30	35	40
4500-6000	35	40	50
6000-10000	40	50	60

Table 13.9 : Representative open area of well screen for various screen sizes

Nominal screen size, inches	Intake area /ft length of screen, sq inches				
	Slot size				
	10	20	40	60	100
3	10	19	32	42	55
10	36	65	110	143	166
20	47	88	156	209	252
30	57	108	192	268	379

Screen size is designated by the size of steel pipe through which screen may be set. Slot number denotes width of openings in thousandths of an inch.

(iii) Well Depth

Expected depth of a well can be determined from well log of the place where well will be dug or from well log of a nearby well. It is desirable to have full penetration of aquifer so that specific capacity would be higher. Under such a case, the well is called as a fully penetrating well.

However if the available screen length is not large enough to fully penetrate the aquifer, it should be centered between top and bottom of the aquifer. Also when the lower portion of the aquifer contains poor quality of water, well screen should be deeper to avoid pumping poor quality water.

(iv) Length of Well Screen

Optimum length of well screen depends on thickness of aquifer available, draw down and nature of aquifer materials as described below.

(a) Homogeneous Artesian/Confined Aquifer

In this case maximum draw down is the distance from initial piezometric level to top of confined aquifer. Screen length of the aquifer depends on the thickness of the aquifer as provided in Table 13.10 (Das Gupta, 1987).

Table 13.10 : Length of screen as varies with thickness of aquifer

Thickness of aquifer	Length of screen
<7.5 m	70% of aquifer thickness
<7.5 - 15 m	75% of aquifer thickness
>15 m	80% of aquifer thickness

The best result is obtained by centering the screen section in the aquifer. However to obtain high yield, screen length or strainer may be provided in the entire thickness of confined aquifer.

(b) Non Homogeneous Artesian Aquifer

Most permeable section of the aquifer should be screened.

(c) Homogeneous Unconfined / Water Table Aquifer

To obtain higher specific capacity, a screen as long as possible may be used. It is preferred to use screen in the bottom one third length. In some cases, bottom half may be screened.

(d) Non Homogeneous Unconfined Aquifer

Principles of design are same as case (ii). i.e. non homogeneous artesian aquifer except that screen should be located in the lower most portion of permeable section.

(v) Well Screen Slot Opening

For homogeneous aquifer slot sizes that contain 40-50% sand are preferable. For non-corrosive wtaer, select higher size opening and for corrosive water, select lower opening. For homogeneous aquifer consisting of gravels and sand, slot openings that retain 30-50% aquifer materials should be selected.

(vi) Well Screen Diameter

Diameter of the intake portion of the well can be varied without affecting specific capacity of well yield. Doubling the diameter of intake section can increase the well yield by about 10%. Screen length depends on thickness of water bearing strata. Screen opening depends on gradation of sand, where as screen diameter depends on factor that total area of screen opening must be such that the entrance velocity of water is 3-6 cm/sec. If velocity is 3-6 cm/sec, then friction loss in screen opening will be less, rate of incrustation will be minimum and rate of corrosion will be minimum.

(vii) Gravel Pack Design

The artificially gravel packed well differs from naturally developed well in the sense that the zone immediately surrounding the well screen is made more permeable by removing the formation materials and then packing with artificially graded coarse material. By doing so, effective diameter of well increases.

In naturally developed well, slot opening retain 40-50% of aquifer materials and allow 60-50% to enter into well during development. In gravel packed well, a graded gravel is so chosen that it will retain all formation materials; a well screen opening is then chosen so that it will retain all gravels. A gravel packed well costs more than a naturally developed well. Artificial gravel pack is necessary for formation like fine uniform sand, thick artesian aquifer, loosely cemented sandstone etc. In general artificial gravel pack is needed when aquifer material is homogeneous with uniformity coefficient <3 and effective grain size <0.25 mm.

Design of gravel pack of well consists of the following steps (Anonymous, 1966; Das Gupta, 1987; Ellithorpe, 1970):

(i) Prepare sieve analysis curves for all strata of aquifer. Determine the stratum composed of the finest sand and select the grading of gravel pack on the basis of sieve analysis.

(ii) Multiply 70% size of sand by a factor in between 4-6: 4 If formation is fine and uniform and 6 if it is coarse and non uniform. If formation sand has highly non-uniform gradation and include silt, then multiplier may be 6-9. If 70% sand is retained by a mesh of 0.013 cm, then multiply 0.013 with 5. Hence D_{70} = 0.065 cm.

(iii) Construct grading curve for gravel pack by trial and error. Through initial point on gravel pack curve and D_{70}, draw a smooth curve representing a material with a uniformity coefficient of 2.5 or less.

Uniformity coefficient is expressed as D_{40}/D_{90}, where D_{40} and D_{90} represents the diameter with 40% and 90% retention.

(iv) Prepare specification of gravel pack material by selecting 4 or 5 sieve sizes that cover the spread of curve.

(v) As a final step, select a size of well screen size openings that retains 90% of gravel pack. Width of slot ranges from 1.5-4.0 mm and length from 5-12.5 m. Thickness of gravel pack $\approx$ 10-20 cm.

13.2.7.7 Common Hydraulic Problems

After a tubewell is constructed and used for some years, there may occur some problems including yielding of low yield. If the problems continue than after some years the well may completely fail to give any yield. The common well problems occur due to (Edward, 1966; Michael, 1978):

(i) Inflow of fine sand, silt and /or muddy water into the tube well.

(ii) Low discharge due to improper well development

(iii) Low discharge and muddy water due to improper gravel pack design and screen

(iv) High rate of pumping which result in the depletion of water in the aquifer

(v) Corrosion or incrustation problem

(vi) Mechanical failure while lowering and setting tube well assembly

(vii) Well screen not provided in good aquifer as per litholog study

(viii) Interference of well

(ix) Wells constructed nearer to barriers like mountain.

13.3. Lift Irrigation

Sometimes water is present in rivers, nallas or ponds but the crop field is situated at a higher elevation than the water level in the sources. In such cases, we to irrigate the crop fields. Canal irrigation is based on gravity flow and so in this case it is not possible to irrigate the crop fields by canals. Tubewell may or may not be feasible there also. Under this context, it is possible to irrigate the crop fields by lifting water through a pump or any indigenous water lifting device. Lifting water through indigenous devices is an age old practice and is age old as the inception of irrigation. It has started several thousands of years ago not only in India but also in many countries in the world. Lifting water through the indigenous devices may be done by manual or animal driven methods. However, the manual driven methods though is very easy yet it is very time consuming as well costly methods compared to animal driven and fuel or electricity driven pumps. With the invent of power supplies especially electricity, it is now easy to irrigate several fields in a small time. Electricity driven pumps are the least costly methods than any other methods to irrigate the crops. Several types of pumps of different makes and models are now available in the markets. Some of these pumps can lift water from low heads such as water in the ponds/nallas etc whereas deep turbine and submersible pumps are preferable to lift water from deep tubewells. Details of different water lifting devices and some information on commonly used pumps are described already in Chapter X.

13.4 Pitcher Irrigation

13.4.1 Introduction

The most common problem in tropical and sub tropical countries like India in the last few decades has been lack of water due to recurrent droughts and uncertainty of rainfall. This has brought about food insecurity at household level as a result of low food and fiber production to meet the demands of the ever increasing population. It is therefore prudent for researchers and scientists to think on conservation and efficient utilization of water to maximize the production per unit volume of the limited water resources. Localized irrigation, which causes wetting of only part of the soil at the base of the plant, is one of the major innovations towards meeting this end. Trickle or drip irrigation and porous tube irrigation are the latest developments, but due to high initial cost few farmers can afford for those. This leads to intensive research on indigenous methods of localized irrigation systems. In view of this, need to develop cheap alternative localized method of irrigation have been felt by the researchers. To achieve this strategy, bamboo drip irrigation for the eastern region, suction irrigation in Rajasthan (Yadav, 1986) and pitcher irrigation at Karnal (Mondal, 1974) have been developed. In the recent past, a number of investigations have been made to develop pitcher irrigation into a scientific irrigation technique by evaluating hydraulics of flow and various management alternatives (Dubey et al, 1987; Mondal et al, 1987[a]). The major conclusion after preliminary and scientific investigations is that this method of irrigation is a practical alternative to drip irrigation and is most efficient in terms of water use efficiency.

Clay pitcher is a simple but efficient technological device, which entails creation of rural employment, savings of scarce water resources, maximizing yield and a reduction in irrigation frequency. For the practical application of pitcher irrigation there is need to have basic information and understanding of system which leads to design criteria for pitcher irrigation. Here water is directly delivered to the plant root zones, in stead of border areas of the field. With pitcher irrigation, deep percolation losses are negligible since water is released from smaller areas, and the rate of water loss can be controlled site to site by the amount of water put in each pitcher (Nirmale et al, 2007). Water requirement in a pitcher irrigated field can be even less than those of a drip irrigated system of the same scale due to the very low hydraulic conductivity of the pitchers, as well as reduced evaporation loss.

13.4.2 Description of the System

In its simplest form, pitcher irrigation entails burying an unglazed and porous clay pot next to a seedling or plant. Water poured into pot seeps slowly into the soil, feeding the seedling roots with a steady supply of moisture. Buried clay pot irrigation, which is very commonly known as pitcher irrigation has been used to grow a wide range of annual and perennial plants in China, Pakistan, India, Iran, Mexico and Brazil. In fact, it is even recorded in Chinese texts dating back more than 2000 years. Pitcher irrigation is an ancient irrigation system which seems to have originated in northern Africa and Iran. The system consists of burying the pitchers in the soil up to its neck. When the pitcher is filled up with water, lateral spread of water comes out through the natural pores of the body wall which wets the surrounding soil of the pitcher and creates moisture conditions conducive to the growth of plants (Fig. 13.13). The loss of water from the pitcher is continuously replenished by adding water in the pitcher. The system thus maintains continuous supply of water and, hence can be classified as high frequency, low depth irrigation method. Crops such as pumpkin, bottle-gourd, water melon, musk melon, tomato, radish, onion, cauliflower and many others can be grown by using this irrigation method.

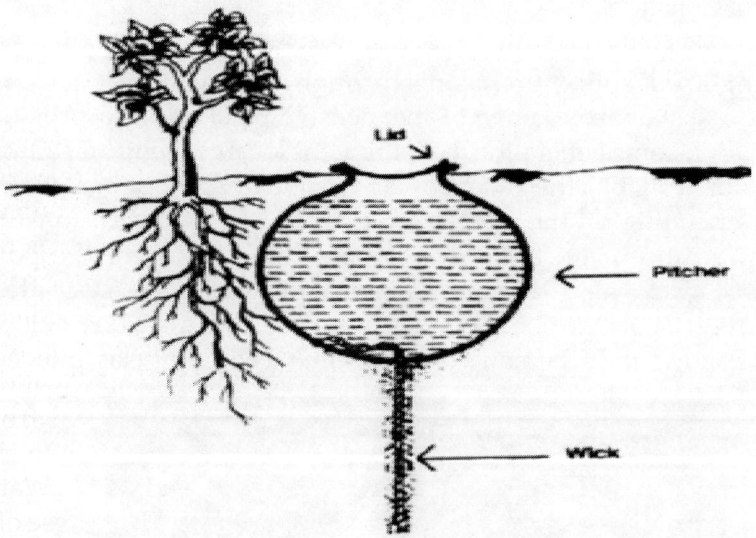

Fig. 13.13 : Buried pitcher and the root system of the tree

13.4.3 Suitability of the System

Pitcher irrigation can be used most effectively under the following conditions.

i) Water is either scarce or expensive

ii) Soils are difficult to level such as undulating terrain

iii) Water is saline and cannot be normally used in surface methods of irrigation

iv) In remote areas where transport of vegetables is expensive and uneconomical

v) When soils are very light with high infiltration rate which can not be successfully irrigated by surface method of irrigation

vi) Availability of enough family labour.

The potential of pitcher irrigations could be exploited over large areas in the state of Orissa where there are prevalent saline zones, undulating hill terrains and where availability of unemployed and/or under-employed family labor exists (Sahu, 1995). From time immemorial there are skilled potters in our state. They mostly prepare baked earthen pitchers and various kitchen utensils.

13.4.4 Pot Construction

Most third world countries possess the knowledge of making and firing unglazed pots. Purchasing through local vendors not only provides a sustainable source of inexpensive pots, but also provides an outlet for income generation, benefiting the local economy. If suitable pots are not available, these can be easily made by hand or in a pottery wheel. Depending on the clay, sand, rice husk or saw dust must be added at a ratio of up to 1:4 to increase the porosity of the pots. Although closed oven firing at temperatures exceeding 450^{0}C is ideal, pits can be fired in open pits at a temperature of 200 to 300^{0}C. Fig. 13.14 shows the schematic view of some well burnt clay pitchers.

Fig.13.14 : Well burnt clay pitchers

13.4.5. Hydraulics of Flow

When a pot filled with water and covered by a lid, is buried in the soil, the water oozes out of the clay pot due to hydraulic head difference between the pot surface and the surrounding soil until it is in equilibrium with the surrounding area. The rate of seepage of water from pitcher depends on the type of plant and soil and climatic conditions around the pot. The movement of water is as a result of the uptake by the crops and it continues as long as the plants take it up and it evaporates. When the surrounding areas become saturated with water and the pot is emptied, water will tend to move back to fill up the pot. The system is therefore self-regulating.

The wetted surface area around the pitcher decides the crop geometry i.e. spacing of plants. The diameter and hence the total wetted area will determine the number of plants which can be accommodated in the wetted zone, where moisture is freely available to plants. The distance of one pitcher from another will influence the wetting pattern. In general pitchers are placed at distances so that wetting fronts do not overlap. The crop characteristics also influence the number of pitchers to be installed in a given area. The wetted diameter and total wetted surface area around a pitcher depends on the texture of the soil and quality of water. For sandy loam soil, values of radius of wetted area and area of the wetted surface around a pitcher for different values of water quality ranging from 0.4 to 15.0 dS/m are reported in Table 13.11 (Mondal et al., 1987[a]).

Table 13.11 : Radius of wetted area and wetted surface area around pitchers

Soil type	Water quality (dS/m)	Radius of wetted area (cm)	Wetted surface area (cm^2)
Sandy loam	0.4	48	7100
	4.0	54	9030
	8.0	57	10075
	12.0	61	11560
	15.0	66	13555

13.4.6 Number and Size of Pitchers

Pitchers varying in capacity from half a litre to more than 15 liters were used in research studies undertaken by Mondal (1974), and Mane et al. (1986). The large capacity pitchers of 7-10 liters were used to grow most vegetable crops such as papaya, ber, grapes and guava. The number of pitchers per ha depended upon the type of the crop grown. The plants with creeping habit normally require larger spacing between the pitchers, while others which form a canopy around the pitcher require less spacing. Therefore number of pitchers per hectare of the field in the former case is worked out to be 2000-2500 while in the latter case it is 4000-5000. The number of pitchers should be calculated by the formula as given below.

$$Number\ of\ pitchers = \frac{Total\ field\ area}{Wetted\ surface\ area\ of\ pitcher} \tag{13.16}$$

13.4.7. Installation of Pitchers

Mondal et al., (1987[b]) reported about the procedure to install the pitcher in the field as follows:

i) For the crop to be grown, select the number of pitchers and locate the places where pitchers are to be buried. Normally, a square grid should be preferred.

ii) At each location, dig a circular pit which is at least 60 cm deep and has a diameter of 90 cm. Preserve the soil brought out from the pit.

iii) Break the soil clods to less than 1 cm diameter and mix enough farm yard manure and basal dose of fertilizers (phosphorus and potash). Also add amendment at this stage, if needed.

iv) Place this mixture of the soil in the pit to give at least 30 cm depth. Place the unglazed pitcher in the centre of the pit (Fig. 13.15).

v) Fill the pitcher with clean water. Avoid muddy water as it may clog the pores.

vi) Sow at least 6-8 seeds/seedlings uniformly around the pitcher. This process is generally carried out two days after the first filling of the pitchers with water. The ideal location of the seedbed is just above the farthest boundary of the pitcher wall.

vii) To facilitate germination pre-soaked and germinated seeds may be sown or water may be added from outside to wet the seedbed. Start filling the pitcher with water at predetermined intervals.

Fig. 13.15 : Installation of pitcher in the field

13.4.8 Movement of Wetting Front and Moisture Distribution

Dubey et al. (1991) conducted experiments on wetting front advance and moisture distribution around the pitcher. The wetting front was observed to advance both horizontally and vertically. The vertical movement was observed to be faster than the horizontal one. Horizontal movement was however restricted and wetting front stopped after some distance. The wetting front advance was represented by a power form equation expressed as:

$$L = a\, t^{b} \qquad (13.17)$$

where L is the length of advancement of water from the centre of the pitcher in cm, t is the time in days and a and b are empirical constants.

The values of empirical constants a and b were found to depend upon the evaporative demand, soil type, discharge of the pitcher and the quality (salinity) of the irrigation water.

The moisture content in the soil around a pitcher was found to vary from a maximum of 20% near the pitcher wall to a normal level of the dry soil, i.e. 5-10% at the end of the wetting front. The maximum moisture was equal to the field capacity of the soil. Thus the soil moisture in the root zone was quite favorable compared to any surface method of irrigation. Salinity of the water did not have much effect on the moisture distribution pattern.

13.4.9 Water Requirement by Pitcher Irrigation

The water requirement of a crop at the farm level served by pitcher irrigation is a function of the type of the crop, soil environment conditions and the attainable on-farm irrigation efficiencies. In general the pitcher irrigation saves water compared to other irrigation methods as described below.

i) Only localized portions of the whole area are irrigated.

ii) Surface evaporation is reduced.

iii) Deep percolation losses are negligible. As a comparison, as much as 50 percent of the water applied in a surface irrigation system goes as deep percolation loss.

iv) It eliminates irrigation runoff.

Mondal et al (1987[a]) conducted experiments and found that the water requirement per pitcher varies widely depending upon the climatic conditions. Under low evaporative demands 1-1.5 litres/day of water is sufficient to replenish the lost water, whereas during high evaporative demands it increases to 2-2.5 litres/day. The quantity of total water needed to raise a crop is however, small. A simple calculation is made to work out total water requirement in cm/ha for varying number of pitchers assuming the maximum replenishment rate 2.5 liters/day (Table 13.12). The method saves large quantities of irrigation water (60-90 percent).

Table 13.12 : Calculation of irrigation water requirement in pitcher irrigation

Number of pitchers	Capacity of pitchers (lit)	Water requirement to fill the pitchers initially (lit)	Water needed for replenishment @ 2.5 lit/day/pitcher for 100 days (lit)	Water requirements (cm/ha)
800	10	8000	20000	2.8
2000	10	20000	50000	5.2
3000	10	30000	75000	7.5
4000	10	40000	100000	10.4

13.4.10 Crop Performance by Pitcher Irrigation

Mondal et al (1987[a]) conducted experiment on crop performance by using pitcher irrigation. The most useful application of this technology could be for growing vegetables and few selected horticultural crops. Some of the important crops which had been successfully grown with this technique at the Central Soil Salinity Research Institute, Karnal were pumpkin, bottle gourd, water melon, musk melon, knol-khol, tomato, bitter gourd, radish, spinach, onion, cauliflower, brinjal, cucumber, sponge gourd and ridge gourd. Performance of all these crops had been good to excellent. Yields of several crops with fresh water were as given in Table 13.13.

Table 13.13 : Yield of field crops irrigated by pitcher irrigation

Crop	Yield (kg/pitcher)	Crop	Yield (kg/pitcher)
Watermelon	11.30	Tomato	5.8
Muskmelon	7.4	Cauliflower	5.2
Bottle-gourd	21.5	Brinjal	5.1
Bitter-gourd	7.5	Cabbage	4.8
Ridge-gourd	4.5	Radish	8.0
Cucumber	14.0	Grapes	3.5

13.4.11 Life Span of the Pitcher

Dubey et al. (1991) reported about the life span of the pitcher. The pitchers as old as 5 years could be used successfully for irrigation purpose. In spite of this they recommended that economic life of the pitchers should be taken as 6 seasons or 3 years. Following precautions in general would increase the life of the pitcher.

i) During irrigation, as for as possible keep the mouth closed as it will prevent sun light to enter into the pitcher and minimize algae formation and growth

ii) Only clean water should be used for filling the pitchers. Runoff from rainfall can be used after passing it through a sand filter.

iii) Before storing the pitchers, ensure proper drying.

13.4.12. Cost Economics

Pitcher irrigation is ideal for sandy to loamy soil with good porosity (40-60%) and for small farmers. Its cost is not more than Rs 4500 per acre, which is about 82% cheaper than drip and the sprinkler irrigation method. Mondal et al. (1987[a]) estimated the economics of irrigation system for different crops. It may vary from crop to crop with different spacing. Total cost is computed as the sum of the fixed cost and the variable cost, which is operational cost. For this irrigation system with 2500 pitchers per hectare of field covered with water melon it is worked out to be Rs. 25000/-. The variable cost of a season is worked out to be Rs. 9640/-. This does not include the cost of cultivation of the crop. The benefit cost ratio of the system is worked out to be 2.5:1. Most of the horticultural crops and vegetables are likely to yield more benefits resulting more benefit cost ratio value, although the quantum of benefits would vary from crop to crop. The corn grown in Mexico on pitcher irrigation showed that the crop production was much higher than that with conventional irrigation methods. Productivity in kg per plant of corn per m^3 of water served by pitcher irrigation and other methods of irrigation is worked out and presented as below (Table 13.14).

Table 13.14 Productivity of corn served by different methods of irrigation

Methods of irrigation	Productivity (kg/plant/m^3 of water)
Closed furrow (basin)	0.7
Sprinkler	0.9
Drip	1.4
Porous capsule (with pressure)	1.9
Deep pipe	2.4
Porous capsule (no pressure)	2.5
Buried clay pot	2.5- 6.0

Soluble fertilizers can also be mixed with water and applied through the pitcher. If the water used for irrigation has high salinity, the pitcher location should be changed every three years. To increase the depth of irrigation, a wick can be added to the pitcher as shown in the figure (The hole in the bottom should be done before firing). The cotton wick must be firmly fixed in the hole to prevent plant roots from entering. If pitchers are used for starting saplings, they can be removed after 1 or 2 years and used elsewhere, because the young trees will have developed deep roots.

Question Banks

1. What is a direct irrigation scheme?
2. What is a storage irrigation scheme?
3. What are the differences between alluvial and non-alluvial canals?
4. Describe the three types of alignment of canals?
5. Describe the distribution systems of canal network by which water is delivered from head regulator to crop fields?

6. What cares should be taken in a canal which has bends in its crops section?
7. Write short notes on
   (i) Gross command area
   (ii) Cultivable command area
   (iii) Intensity of irrigation
   (iv) Duty and
   (v) Delta.
8. What is difference between duty and delta ?
9. Describe a relationship between dates and delta.
10. The gross commanded area for a distributary is 6000 hectares, 65% of which is cultivable. The intensity of irrigation for *rabi* season is 50% and that for *kharif* season is 40%. If the average duty at the head of the distributary is 1500 hectares/cumec for *rabi* season and 1000 hectares/cumec for *kharif* season, find out the discharge required at the head of the distributary from average demand considerations.
11. The gross command area for a distributary is 5000 hectares, 50% of which is cultivable. The intensity of irrigation for *rabi* season is 60% and that for *kharif* season is 70%. Groundnut is grown in the entire command in *rabi* and paddy in *kharif*. Find out the discharge required at the head of the distributary considering the peak water requirement of *rabi* crop 20 cm and the period of peak water requirement 3 weeks and that for *kharif* cop 25 cm with peak period of water requirement 30 days.
12. The cultivable commanded area of watercourse is 1400 hectares. Intensities of irrigation of sugarcane and groundnut crops grown in *rabi* season are 35% and 45%, respectively. The duties for the crops at the head of the watercourse are 850 hectares/cumec and 1450 hectares/cumec, respectively. Compute the discharge required at the head of the watercourse.
13. What is a regime channel? Describe the design of regime channel by Kennedy's and Lacey's theory.
14. Write drain the various types of design of a regime irrigation channel.

15. An irrigation channel has side slope 2:1 and base width 5 m. It has flow depth of 1.2 m. The bed slope of the channel is 1:2500. Assume Manning's rugosity coefficient n = 0.026. Determine if the channel will be regime channel. Assume critical velocity ratio as 1.0.
16. An economic rectangular channel section is carrying a discharge of 1.5 cumecs. Water carried by the channel contains some silt. The value of Manning's roughness coefficient of the channel section is 0.025. Use the velocity of silt laden water in the channel by Kenned's theory as $V = 0.55mD^{0.64}$. The value of m (critical velocity ratio) is assumed as 1.0. Find out the bed slope of the channel to carry the design discharge of 1.5 cumecs.
17. Why lining of irrigation channels are necessary?
18. What are the advantages of lining the irrigation channels?
19. Describe how lining of irrigation channel help in elimination of waterlogging and salinity in irrigation command.
20. Discuss the construction practices of different types of lining.
21. Why are the repair and maintenance of irrigation channel required ?
22. What is a well irrigation? What is its advantages and disadvantages?
23. What points need to be taken into consideration for selection of ideal site of design of well.
24. What is an open well? Write notes on various types of open well?
25. Differentiate between impervious and previous lined well.
26. What are the differences between shallow and deep well.
27. Describe the pumping test and recuperation test for determining the yield of an open well.
28. Design an open well in fine sandy soil with safe yield 0.45 m^3/hr /m^2/ unit depression head so that an yield of 0.0045 m^3/sec is discharged from well with depression head h = 3 m
29. Calculate the average yield of a well of 2,5 m diameter by recuperation test. The water level in the well is decreased to an extent of 3.0 m. The rate of rise of water is 1.0 m/hr. Assume average depression head, h =3 m.

30. What is tubewell irrigation? Describe the various advantages and disadvantages of tubewell irrigation.
31. Write short notes on various types tubewells.
32. Draw a typical discharge drawdown curve of shallow and deep tubewells and explain how the discharge varies with types of tubewells for a given discharge.
33. Describe in details about the various design parameters of tubewells.
34. With a label diagram describe about the structure of a tubewell.
35. What are the factors on which optimum length of well screen depend on ?
36. What is difference between naturally developed well and artificial gravel packed well? Narrate the different design procedure of gravel packed well.
37. What are the common hydraulic problems of a tubewells ?
38. What is lift irrigation? How water is lifted to crop fields situated at higher elevation from rivers and ponds.
39. Pitcher irrigation is sometimes called as clay pot irrigation. Justify?
40. What is pitcher irrigation? What are its advantages.
41. Describe in short the description of pitcher irrigation in crop fields.
42. Write notes on suitability of pitcher irrigation system.
43. How is clay pots constructed?
44. How can you determine the number and sizes of pitchers?
45. Write the different procedures for installation of pitchers in fields.
46. Describe the wetting front movement and moisture distribution pattern of pitcher irrigation system.
47. How can you determine the water requirement of pitcher irrigation?
48. How does the pitcher irrigation influence the crop production and productivity.
49. What procedures need to be undertaken for increasing the life span of clay pots in pitcher irrigation?
50. Describe how pitcher irrigation is economically variable for small and marginal farmers.

References

Ahrens, T. P. 1958. Water Well Engineering. Water Well Jl. 12, December.

Anonymous. 1966. Procedure used in gravel packing wells. Johnsons Drillers Jl. 38 : 2.

Anonymous. 1994. Annual Report of AICRP on Water Management, Chiplima Centre, Orissa University of Agriculture and Technology, Bhubaneswar, Orissa, pp. 105.

Benison, E. W. 1947. Ground Water. E. E. Johnson Inc., St. Paul, Minn. 500p.

Campbell, M. D. and Lehr, J. H. 1973. Water Well Technology. Mc-Graw-Hill Book Co., New York p. 681.

Das Gupta, A. 1987. Handouts on Ground Water Hydrology, Asian Institute of Technology, Bangkok, pp. 282.

Dubey, S.K., Gupta, S.K. and Mondal, R.C. 1991. Pitcher irrigation techniques for arid and semi-arid zone, Dry Land Research and Technology, 6 (D-7), 137-177.

Dubey, S.K., Mandal, R.C. and Gupta, S.K.1987. Annual Reports, C.S.S.R.I, Karnal, India, 54-55.

Dylla, A.S. and Muckel, D. C. 1967. Experimental development of shallow ground water wells. Pub. University Nevada and U. S. Dept. of Agr. 23 p.

Edward, E. Johnson Inc. 1966. Ground Water and Wells. Pub. Edward E. Johnson Inc., Saint Paul, Minnesota, U. S. A., 440 p.

Ellithorpe, G. L. 1970. Current practices relative to the design and placement of artificial gravel pack for tubewells. Proc. Summer Institute on Tube Well Technology I.A.R.I., New Delhi.

Garg, S.K. 1987. Irrigation Engineering and Hydraulic Structures. Khanna Publishers, Delhi, pp. 1184.

Mane, T.A., Khade, K.K., Pampattiwar, P.S. and Unde, P.H. 1986. Proc. of Int. Seminar, Water Management in Arid and Semi Arid Zones, Hissar, India, 558-566.

Michael, A. M. and Ojha, T. P. 1966. Principles of Agricultural Engineering, Vol. II. Pub. Jain Brothers, Jodhpur. 534 p.

Michael, A.M. 1978. Irrigation Theory and Practices. Vikash Publishing House Pvt. Ltd., New Delhi.

Michael, A.M., Arora, D. R., Bhattacharya, A. K., Mandal, A. and Gupta, A. K. 1970. Handbook of Farm Irrigation Structures. Indian Agricultural Research Institute, New Delhi. 82 p

Mondal, R.C. 1974. World crops 26: 94-97.

Mondal, R.C., Dubey, S.K. and Gupta, S.K. 1987[a]. Annual Report, Central Soil Salinity Res. Institute, Karnal, India.

Mondal, R.C., Gupta, S.K. and Dubey, S.K.1987[b]. Pitcher Farming, Extension bulletin, C.S.S.R.I., Karnal-132001.

Nirmale, A., Kumar, V., Kothari, M., Purohit, R.C. and Singhvi, B.S.M. 2007. Hydraulics study of earthen pitchers, Journal of Agricultural Engineering, Vol, 44(2): April-June.

Sahu, G.S. 1995. Hydraulics and thermal behavior of irrigating pitcher, unpublished M.Tech. Thesis, Dept. of Soil and Water Conservation Engineering, CAET, OUAT, Bhubaneswar.

Sharma, R.K. 1987. A Text Book of Hydrology and Water Resources. Dhanpat Rai and Sons, Delhi, pp. 818.

Sharma, R.K. and Sharma, T.K. 2002. Irrigation Enginering Including Hydrology. S. Chand and Company Ltd., New Delhi.

Sivanappan, R.K. 2000. Micro irrigation including low energy precision application. International Conference on Managing Natural Resources for Sustainable Agricultural Production in the 21st Century. Organised by Indian Society of Soil Science, New Delhi: 231-234.

Slichter, C. S. 1989. Theoretical investigation of the motion of ground-waters. 19th Annual Report, U.S.G.S., 295-384.

Yadav, R.K. 1986. Intensive Agriculture, 24: 18-26.

Conversion Tables

Unit	Multiply by	To obtain
Acre	4.05×10^3	Square meter, m^2
Acre	0.405	Hectare, ha ($10^4 m^2$)
Acre	4.05×10^{-3}	Square kilometer, km^2
Angstrom	0.1	Nanometer, nm (10^{-9} m)
Atmosphere	0.101	Mega pascal, M pa (10^6 Pa)
Bar	10^{-1}	Mega pascal, M pa (10^6 Pa)
British thermal unit	1.05×10^2	Joule, J
Calorie	4.19	Joule, J
Calorie per square centimeter / minute	698	Watt per square meter, W m^{-2}
Centimeter per day (elongation rate)	0.116	Micrometer per second, μms^{-1} ($10^{-6} ms^{-1}$)
Cubic feet	0.028	Cubic meter, m^3
Cubic inch	1.64×10^{-5}	Cubic meter, m^3

Cubic inch	16.4	Cubic centimeter, cm^3 (10^{-6} m^3)
Curie	3.7 x 10^{10}	Becquerel, Bq
Degrees (angle)	1.75 x 10^{-2}	Radian , rad
Dyne	10^{-5}	Newton, N
Erg	10^{-7}	Joule, J
Foot	0.305	Metre,m
Foot-pound	1.36	Joule, J
Bar	10^3	Kilo pascal, K pa
Gallon (U.S)	3.78	Litre, L (10^{-3} m^3)
U.S. gallon per acre	9.35	Liere per hectare, 1it ha^{-1}
Gram per cubic centimeter	1.00	Mega gram per cubic metre, Mg m^{-3}
Inch	25.4	Millimeter, mm (10^{-3} m)
Micromole (H_2O) per square centimeter second (transpiration)	180	Milligram (H_2O) per square metre per second, mg m^{-2} s^{-1} (10^{-3} g m^{-2} s^{-1})
Micron	1.00	Micro metre, μm (10^{-6} m)
Mile	1.61	Kilometer, km (10^3 m)
Miles per hour	0.477	Metre per second, m s^{-1}
Milli mhos per centimeter	1.00	Deci siemens per metre, ds m^{-1}
Ounce	28.4	Gram, g (10^{-3} kg)
Ounce (FLUID)	2.96 x 10-2	Litre, (10^{-3} m^3)
Pint (fluid)	0.473	Litre, (10^{-3} m^3)
Pound	454	Gram, g (10^{-3} kg)
Pound per acre	1.12	Kilogram per hectare, kg ha^{-1}
Pound per acre	1.12 x 10-3	Mega gram per hectare, Mg ha^{-1}
Pound per cubic feet	16.02	Kilogram per cubic metre, kg m^{-}
Pound per cubic inch	2.77 x 104	Kilogram per cubic metre, kg m^{-3}
Pound per square feet	47.9	Pascal, pa
Pound per square inch	6.9 x 103	Pascal, pa
Quart (liquid)	0.946	Litre, (10^{-3} m^3)
Quintal (metric)	102	Kilogram, kg

Square centimeter per gram	0.1	Square metre per kilogram, $m^2 kg^{-1}$
Square feet	9.29 x 10-2	Square metre, m^2
Square inch	645	Square millimeter, mm^2 (10^{-6} m^2)
Square mile	2.59	Square kilometer, km^2
Square millimeter per gram	10^{-3}	Square metre per kilogram, $m^2 kg^{-1}$
Temperature (^{0}F-32)	0.556	Temperature, ^{0}C
Temperature (^{0}C+273)	1	Temperature, K
Ton (metric)	10^3	Kilogram, kg
Ton (2,000 lb)	907	Kilogram, kg
Ton (2,000 lb) per acre	2.24	Mega gram per hectare, Mg ha^{-1}

Conversion Factors for Different Units of Length

1 metre (m)	3.208 feet (ft)
1 metre (m)	39.37 inches (in)
1 centimetre (cm)	0.3937 inch
1 kilometre (km)	1,000 metres (m)
1 kilometre (km)	0.621 mile (statute mile, mi) = 0.5396 nautical mile (n mi)
1 foot	0.305 metre
1 inch (in)	2.54 centimetre
1 mile (mi)	5,280 feet
1 mile (mi)	1609 kilometre
1 micrometre (μm.)	10^{-6} m
1 angstrom (^{0}A)	10^{-10} m
1 metre	1.09361 yard
1 kilometre	1093.61 yard
1 inch	0.02778 yard
1 foot	0.33333 yard
1 mile	1760 yard

Conversion Factors for Different Units of Area

1 square metre	10.764 square feet
1 square centimetre	0.155 square inch
1 square kilometre	100 hectares
1 square kilometre	0.3861 square mile
1 hectare	10,000 square meres
1 hectare	107,640 square feet
1 hectare	2.741 acres
1 square foot	0.0929 square metre
1 square inch	6.452 square centimeter
1 acre	43,560 square feet
1 acre	0.4047 hectare
1 square mile	640 acres
1 square mile	258.99 hectares
1 square mile	2.59 square kilometres
1 square metre	1.19599 square yard
1 acre	119.599 square yard
1 hectare	11960 square yard
1 acre	4840 square yard
1 hectare	100 are
1 square metre	0.01 are
1 square kilometre	10000 are

Conversion Factors for Different Units of Volume

1 cubic metre	35.314 cubic feet
1 cubic metre	1.308 cubic yards
1 cubic metre	1,000 litres
1 cubic metre	106 cubic centimeters (cm^3)
1 litre	0.035 cubic feet
1 litre	0.2642 U.S. gallon (gal.)
1 litre	0.2201 imperial gallon
1 litre	10^{-3} cubic metre (m^3)
1 cubic centimetre	0.061 cubic inch

1 cubic foot	0.0283 cubic metre
1 cubic foot	23.32 litres
1 cubic foot	7.4 U.S. gallons
1 cubic foot	6.23 imperial gallons
1 cubic inch	16.39 cubic centimeters
1 cubic yard	0.7645 cubic metre
1 U.S. gallon	3.7854 litres
1 U.S .gallon	0.833 imperial gallon
1 imperial gallon	1.201 U.S. gallon
1 imperial gallon	4.5436 litres
1 acre foot	43,560 cubic feet
1 acre foot	1233.5 cubic metres
1 acre inch	3,630 cubic feet
1 acre inch	102.8 cubic metres

Conversion Factors for Different Units of Weight

1 gram (gm)	0.0353 ounce (oz)
1 gram (gm)	0.0022 pound (lb)
1 kg	2.205 pounds (lb)
1 quintal	100 kg
1 metric ton	2,205 pounds (lb)
1 metric ton	1,000 kg
1 metric ton	10 quintals
1 metric ton	1.1 ton
1 kilogram (kg)	1,000 gm
1 ounce	28 gm
1 pound	0.45 kg
1 ton	0.9 metric ton (m.t)

Conversion Factors for Different Units of Pressure

1 bar	10^6 dynes / cm^2
1 bar	1,023 cm of water column
1 atmosphere	1,032 cm of water column

1 atmosphere	76 cm Hg
1 atmosphere	1.013 bar
1 inch Hg	0.0334 atmosphere
1 inch H_2O	2.49 millibar (mb)
1 mbar (mb)	0.75mm Hg
1 lb/in2 (psi)	51.72 mm Hg.
1 atm	14.70 lb/in^2 (psi)
1 atm	29.92 in Hg
1 millibar (mb)	0.0145 lb/in^2 (psi)
1 millibar (mb)	1,000 dynes/cm^2
1 millibar (mb)	1000 newton/m^2 (N/m^2)
1 millibar (mb)	100 Pascal (Pa)

Conversion Factors for Different Units of Velocity

1 m/sec	3.60 km/hr
1 m/sec	2.24 mi/hr
1 m/sec	1.94 knots
1 knots	1 n mi/hr
1 knots	1.15 mi/hr
1 knots	0.515m/sec

Conversion Factors for Different Units of Energy

1 calorie (cal)	4.187 joules (J)
1 calorie	4.187 x 10^7 ergs
1 calorie	1.16 x 10^{-6} killowatt hour (Kwh)
1 calorie	3.97 x 10^{-3} British thermal units (Btu)
1 calorie	1 langely-centimetre^2 (ly-cm^2)

Conversion Factors for Power

1 watt (W)	1 joule / sec
1 watt (W)	0.239 cal/sec
1 watt (W)	0.0569 Btu/min
1 watt (W)	0.00134 electrical horse power
1 watt (W)	2.39 x 10-5 ly-m^2 / sec

Conversion Factors for Radiation to Equivalent Depth Evaporation

1 cal/cm^2	1/59 mm
1 cal/cm^2 /min^{-1}	1 mm/hr
1 cal/cm^2/ min (24 hr)	24 mm / day
1 mw /cm^2	1/70 mm/hr
1 mw/cm^2 (24hr)	0.344 mm/day
1 joule / cm^2 / min (24 hr)	5.73 mm/day

Conversion Factor for Energy Flux Density

1 cal cm^{-2} min^{-1}	1 langely min^{-1} = 69.8 milli watt cm^{-2}
1 mm hr^{-1} (evaporation of water based on latent heat of vaporization of 590 cal g^{-1}	59 cal cm^{-2} hr^{-1} = 68.9 milli watt cm^{-2}

Conversion Factor for Rates of Flow

1 cubic metre per second (m^3 / sec)	35.314 cubic feet per second (cu.ft / sec)
1 cubic metre per hour (m^3/hr)	0.278 litres per second (l/sec)
1 cubic metre per hour	4.403 U.S. gallon per minute
1 cubic metre per hour	3.668 imperial gallons per minute
1 U.S gallon per minute	0.06309 litres per second
1 imperial gallon per minute	0.07573 litre per second
1 litre per second	0.0353 cubic foot per second
1 litre per second	15.852 U. S.gallons per minute
1 litre per second	13.206 imperial gallons per minute
1 litre per second	3.6 cubic metres per hour
1 cubic foot per second	0.0283 cubic metres per second
1 cubic foot per second	28.32 litrs per second
1 cubic foot per second	448.8 U. S. gallons per minute
1 cubic foot per second	373.8 Imperial gallons per minute
1 cubic foot per second	1 acre inch per hour (approx)
1 cubic foot per second	2 acre foot per day (approx)

1 cusec flowing continuously for a day	0.09 million cu.ft =1.98 acre feet =2446.57 cubic metres =0.54 million gallon (imp)
1 cusec flowing continuously for a year	= 724 acre fee =31.53 million cu.ft

Some Useful Approximations

2.5 cm	1 inch (in)
30 cm	1 foot
10 metres	33 ft
10 Km	6 miles
16 Km	10 miles
1 hectare	2.5 acres
1 square inch	6.5 square cm
1 square metre	11 square feet
16.4 cubic centimeters	1 cubic inch
35 cubic feet	1 cu.metre

Miscellaneous

Milli curie (m.Ci)	= The unit of radio activity equivalent to 3.7×10^7 disintegration / second
1 atm. osmotic pressure	0.36 x E.C in m mhos/cm
1 ppm	=0.64 x E.C in m mhos /cm =meq / litre x equivalent weight
1 milli-equivalent per litre	10 x E.C in m mhos/cm
1 ton per acre foot	0.00136 p.p.m
1 meq/100 gm of soil of exchangeable sodium	1.72 tons/acre foot of gypsum requirement =0.32 tons/ acre foot sulphur requirement
1 ton of sulphur	4.17 ton lime sulphur solution =3.06 ton H_2SO_4 =5.38 ton gypsum
ppm	Parts per million
ppm	1 microgram (μg) per ml of water
ppm	1 microgram (μg) per gm

ppm	1 μg per 1,000 mgm
ppm	1 μg per 1000.000 microgram
1 gm	=1000gm =1000 x 1000 μg= 10^6 μg
ml x Normality	Meq (milliequivalent)
Meq/lit x eq. wt	ppm

PPM to Percentage

Ppm %	= Parts per 1,000
1000,000/10 i.e ppm/10,000	percentage
Percentage x 10,000	ppm

Index

C

D

E

F

K

L

M

N

O

P

Q

R

T

U

V

W

Y